T0296882

Topological Data Analysis with Applications

The continued and dramatic rise in the size of data sets has meant that new methods are required to model and analyze them. This timely account introduces topological data analysis (TDA), a method for modeling data by geometric objects, namely graphs and their higher-dimensional versions, simplicial complexes. The authors outline the necessary background material on topology and data philosophy for newcomers, while more complex concepts are highlighted for advanced learners. The book covers all the main TDA techniques, including persistent homology, cohomology, and Mapper. The final section focuses on the diverse applications of TDA, examining a number of case studies ranging from monitoring the progression of infectious diseases to the study of motion capture data.

Mathematicians moving into data science, as well as data scientists or computer scientists seeking to understand this new area, will appreciate this self-contained resource which explains the underlying technology and how it can be used.

Gunnar Carlsson is Professor Emeritus at Stanford University. He received his doctoral degree from Stanford in 1976, and has taught at the University of Chicago, at the University of California, San Diego, at Princeton University, and, since 1991, at Stanford University. His work within mathematics has been concentrated in algebraic topology, and he has spent the last 20 years on the development of topological data analysis. He is also passionate about the transfer of scientific findings to real-world applications, leading him to found the topological data analysis-based company Ayasdi in 2008.

Mikael Vejdemo-Johansson is Assistant Professor in the Department of Mathematics at City University of New York, College of Staten Island. He received his doctoral degree from Friedrich-Schiller-Universität Jena in 2008, and has worked in topological data analysis since his first postdoc with Gunnar Carlsson at Stanford 2008–2011. He is the chair of the steering committee for the Algebraic Topology: Methods, Computation, and Science (ATMCS) conference series and runs the community web resource appliedtopology.org.

Topological Data Analysis with Applications

GUNNAR CARLSSON
Stanford University, California

MIKAEL VEJDEMO-JOHANSSON
City University of New York, College of Staten Island and the Graduate Center

CAMBRIDGE
UNIVERSITY PRESS

University Printing House, Cambridge CB2 8BS, United Kingdom

One Liberty Plaza, 20th Floor, New York, NY 10006, USA

477 Williamstown Road, Port Melbourne, VIC 3207, Australia

314–321, 3rd Floor, Plot 3, Splendor Forum, Jasola District Centre,
New Delhi – 110025, India

103 Penang Road, #05–06/07, Visioncrest Commercial, Singapore 238467

Cambridge University Press is part of the University of Cambridge.

It furthers the University's mission by disseminating knowledge in the pursuit of
education, learning, and research at the highest international levels of excellence.

www.cambridge.org
Information on this title: www.cambridge.org/9781108838658
DOI: 10.1017/9781108975704

First published 2022

A catalogue record for this publication is available from the British Library.

Library of Congress Cataloging-in-Publication Data
Names: Carlsson, G. (Gunnar), 1952– author. | Vejdemo-Johansson, Mikael, 1980– author.
Title: Topological data analysis with applications / Gunnar Carlsson, Mikael Vejdemo-Johansson.
Description: New York : Cambridge University Press, 2021. |
    Includes bibliographical references and index.
Identifiers: LCCN 2021024970 | ISBN 9781108838658 (hardback)
Subjects: LCSH: Topology. | Mathematical analysis. | BISAC: MATHEMATICS / Topology
Classification: LCC QA611 .C29 2021 | DDC 514/.23–dc23
LC record available at https://lccn.loc.gov/2021024970

ISBN 978-1-108-83865-8 Hardback

Contents

Preface

Data sets come in many shapes and sizes. The data sets presented in the figure below illustrate this point very well.

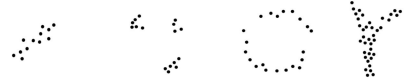

- The data set on the far left has the rough shape of a line in the plane. We are all familiar with many examples of this kind of data, and such data are typically handled with various regression models, which permit prediction and also allow for greater understanding of the data. This helps in developing mental models.
- The second set from the left illustrates a data set which decomposes into disjoint groups, and is not well approximated by any line. This kind of data occurs very frequently in the biomedical and social sciences, and cluster analysis has been developed to produce such decompositions in order to deliver taxonomies for the data.
- The third set from the left is a type of data set that occurs frequently when one is dealing with time series data representing periodic or recurrent behavior of some kind.
- The data set on the far right might describe data in which there are one standard or normal mode and three extremal modes. For example, it might come from sensors on an airliner, where the standard mode is flying at altitude in non-turbulent conditions and where the three extremal modes are takeoff, landing, and flying at altitude in turbulent conditions.

The first two data sets have dedicated methodologies (regression and cluster analysis, respectively) for their analysis. The latter two do not, although we believe that one could develop such methods for each of these two types. However, since these four data sets are by no means a complete list of the possible shapes of data, we can rapidly convince ourselves that hoping to create a complete list of shapes with tailor-made methods for each data shape is not the best solution to the problem dealing with all the different complexities that we can expect to find.

A possible approach to modeling data sets like those above is to view a modeling mechanism as a way to approximate data by sets with various shapes. For example, linear regression is the approximation of data by lines, planes, etc., while cluster analysis can be viewed as the approximation of data by finite discrete sets of points. Using this approach, the third data set from the left in the figure above could be approximated by a loop, and the data set on the far right could be thought of as having the shape of a letter "Y", i.e., of three line segments which all join at a single central point. What these observations suggest is that we should develop a single method that can represent *all* shapes in one package. Fortunately, the mathematical discipline called topology provides exactly such a method. It turns out that graphs (and somewhat more complex objects called simplicial complexes) are very useful ways to describe shapes.

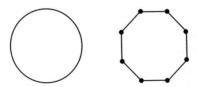

In this figure, we see a circle on the left and an octagon on the right. The two are very similar, in the sense that the octagon approximates the circle well, although it does not capture the curvature. There is a highly non-linear parametrization in which the points of the circle encode points on the octagon, and vice versa, which is called a *homeomorphism*, and we regard the two as representing the same information.

However, the octagon can also be described by a purely combinatorial object, namely a list of its vertices and of its edges, together with the information about which vertex belongs to which edge. This kind of parametrization is called a triangulation of the circle, and is the key concept from topology for the study of data. In particular, we will develop methods for approximating data sets by shapes described in a combinatorial way in the same way as we described the octagon above. The relevant combinatorial objects are graphs (in the computer science or combinatorics sense) or objects called simplicial complexes, which include not only edges but higher-order subsets such as triangles, tetrahedra, and higher-dimensional analogues.

With the above discussion in mind, topological data analysis (TDA) can be summarized as the idea that data, like topological spaces, can be usefully modeled by combinatorial objects such as graphs and simplicial complexes. The subject has been developing rapidly over the last 20 years, and this volume is an attempt to describe the theory as well as a varied array of applications. The rationale for the development of these methods consists of a number of different observations concerning data science.

- Linear algebraic methods, such as principal component analysis or multidimensional scaling, because of their algebraic nature are often not flexible enough to capture complicated non-linearities in data and their scatterplot output is often not as informative as one would like. Simplicial complex models are more flexible and capable of expressing complexity in the data. They also admit a great deal of functionality, allowing for effective interrogation and search of the data, as detailed in our Section 4.3.5.

- Cluster analysis seeks to divide data into disjoint groups to create a taxonomy of the data set. In many situations where cluster analysis is applied, however, one finds that the natural output is not a partition of the data into disjoint groups but, rather, a "soft clustering" in which the data is broken into groups that may overlap. This kind of information is very naturally modeled with a simplicial complex, which is able to describe the relationships between groups implied by the overlaps, using an appropriate shape or space. An ordinary cluster decomposition, by a partition, is in this situation modeled by a zero-dimensional simplicial complex, i.e. a finite set of points.

- Data science is often confronted with the problem of deciding the appropriate shape for a data set, so as to be able to model it effectively. The theory of simplicial complexes is equipped with a method for describing the shape structure of the output of a model; this is based on an extension of the *homology* construction from the algebraic topology of spaces. The extension is called *persistent homology* and will be the subject of many of the ideas that we present.

- Feature selection and feature engineering is a major task in data science. It is particularly challenging for data sets which are unstructured in the sense that they are not well represented by a data matrix or a spreadsheet with numerical entries. For example, a database of large molecules is regarded as unstructured because it consists of an unordered set of atoms and an unordered set of bonds and, further, because the spatial coordinates of the atoms can be varied via a rigid motion of space while the structure of the molecule remains unchanged. This means that a representation by the coordinates of the atoms is not meaningful. It turns out, though, that the molecules themselves have a geometry expressed through inter-atomic distances, which allows us to apply the homology tools described above to generate meaningful numerical quantities that can be used for analysis. Images form another class of data which can be viewed as unstructured and which can be studied with homological methods.

- Another way in which topological methods can be used for feature engineering is the notion of topological signal processing (Robinson 2014). The idea here is that, when given a data matrix, it is also useful to develop a topological model for the columns of the data matrix (i.e. the features) rather than for the points or samples of the data (i.e. the rows). In this way, each of the original data points can be viewed as a function on the set of features and ultimately as a function on the topological model. Incorporating various methods, including graph Laplacians, one can impose structure on data points using this approach, and obtain topologically informed dimensionality reductions.

We believe that the use of TDA in data science will motivate interesting and useful developments within topology. It is therefore useful to see where TDA methods fit within standard algebraic topology and homotopy theory. Here are some important points about this fit.

- Persistent homology can be described as the study of diagrams whose shape is defined by the partially ordered set $\mathbb{R}$. A number of other diagrams have been studied, including those used in zig-zag persistence and multidimensional persistence. As the work in TDA broadens and deepens, it is likely that increasingly sophisticated diagrams will be useful for extracting more detailed information from data sets. It follows that the construction of invariants for diagrams of various shapes will be a useful endeavor.

- Because TDA operates by studying samples of discrete sets of points, the dimensions of spaces that can be analyzed solely by TDA methods are fairly low, for the most part ≤ 5. A data set which would faithfully represent a space of dimension 10 would be expected to require at least 10^{10} points, if one assumes a resolution of 10 points for each dimension. This is already a very large number, and demonstrates the point that, for example, 50-dimensional homology is not likely to occur in a useful way in data sets. This suggests that studying more sophisticated unstable homotopy invariants (such as cup products, Massey products, etc.) would be a good direction to pursue. For example, the use of cup products is a key part of Carlsson & Filippenko (2020).

- Within algebraic topology and homotopy theory, a very interesting aspect is the topology of spaces equipped with a reference map to a base space B; this is referred to as parametrized topology. All maps are then required to respect the reference map. The category of spaces over a base contains a much richer set of invariants than in the absolute case (i.e. ordinary topology, without a reference map), where B is a single point. This idea comes up in the study of evasion problems (Carlsson & Filippenko 2020) and can be used to define the idea of data science over a base, or parametrized topological data analysis (Nelson 2020), which appears to be a useful framework for an iterative method of data analysis. The study of unstable invariants in this case is particularly rich, and warrants further attention.

- One is often interested in studying the invariants of a space X which are not necessarily topological in nature but which nevertheless can be thought of as qualitative, for example, the notion of the corners or ends of spaces are examples of this kind of situation. One way to approach such problems is to perform constructions on X so as to produce an associated space which reflects the property one wants to study, and then to use topological methods such as homology to perform the analysis. A powerful example of this philosophy is the work of Simon Donaldson on the topology of smooth 4-manifolds, where he showed that certain moduli spaces attached to smooth 4-manifolds allow one to study the topology of the manifold itself (Donaldson 1984). This kind of approach can be used to investigate various shape distinction problems that are not directly topological in nature.

The goal of this book is to introduce the ideas of topological data analysis to both data scientists and topologists. We have omitted much of the technical material about topology in general, as well as for homology in particular, with the expectation that the reader who has studied the book will be able to go further in the subject as needed. We hope that it will encourage both groups to participate in this exciting intellectual development.

As a matter of convention, we choose to use the terms injective, surjective, and bijective instead of one-to-one, onto, and "one-to-one and onto".

The authors are very grateful for helpful conversations with many people, including R. Adler, A. Bak, E. Carlsson, J. Carlsson, F. Chazal, J. Curry, V. de Silva, P. Diaconis, H. Edelsbrunner, R. Ghrist, L. Guibas, J. Harer, S. Holmes, M. Lesnick, A. Levine, P. Lum, B. Mann, F. Mémoli, K. Mischaikow, D. Morozov, S. Mukherjee, J. Perea, R. Rabadan, H. Sexton, P. Skraba, G. Singh, R. van de Weijgaert, S. Weinberger, and A. Zomorodian.

We particularly thank A. Blumberg, whose collaboration on early drafts of this book has helped immensely.

Deep thanks also go to the editorial staff at Cambridge University Press, for their patience and all their help.

Part I

Background

1   Introduction

In the last two or three decades, the need for machine learning and artificial intelligence has grown dramatically. As the tasks we undertake become ever more ambitious, both in terms of size and complexity, it is imperative that the available methods keep pace with these demands. A critical component of any such method is the ability to model very large and complex data sets. There is a large suite of powerful modeling methods based on linear algebra and cluster analysis that can often provide solutions for the problems that arise. Although they are often successful, these methods suffer from some weaknesses. In the case of algebraic methods, it is the fact that they are not always flexible enough to model complex data, such as data sets of financial transactions or of surveys. Clustering methods by definition cannot model continuous phenomena. Additionally, they often require choosing thresholds for which there are no good theoretical justifications. What we will discuss in this volume is a modeling methodology called topological data analysis, or TDA, in which data is instead modeled by geometric objects, namely graphs and their higher-dimensional versions, simplicial complexes. Topological data analysis has been under development during the last 20 or so years, and has been applied in many diverse situations. Its starting point is a set equipped with a metric, typically given as a dissimilarity measure on the data points, which can be regarded as endowing the data with a shape. This shape is very informative, in that it describes the overall organization of the data set and therefore enables interrogation of various kinds to take place; TDA provides methods for measuring the shape, in a suitable sense. This is useful in as much as it allows one to access information about the overall organization. In addition, TDA can be used to study data sets of complex description, which might be thought of as unstructured data, where the data points themselves are sets equipped with a dissimilarity measure. For example, one might consider data sets of molecules, where each data point consists of a set of atoms and a set of bonds between those atoms, and use the bonds to construct a metric on the set of atoms. This idea leads to powerful methods for the vectorization of complex unstructured data. The methodology uses and is inspired by the methods of topology, the mathematical study of shape, and we now give a more detailed description of how it works.

Much of mathematics can be characterized as the construction of methods for organizing infinite sets into understandable representations. Euclidean spaces are organized using the notions of vector spaces and affine spaces, which allows one to arrange the (infinite) underlying sets into understandable objects which can readily be manipulated and which can be used to construct new objects from old in systematic ways.

Similarly, the notion of an algebraic variety allows one to work effectively with the zero sets of sets of polynomials in many variables. The notion of shape is similarly encoded by the notion of a *metric space*, a set equipped with a distance function satisfying three simple axioms. This abstract notion permits one to study not only ordinary notions of shape in two and three dimensions but also higher-dimensional analogues, as well as objects like the *p*-adic integers, which may not be immediately recognized as being geometric in character. Thus, the notion of a metric serves as a useful organizing principle for mathematical objects. The approach that we will describe demonstrates that the notion of metric spaces acts as an organizing principle for finite but large data sets as well.

Topology is one of the branches of mathematics which studies properties of shapes. The aspect of the study of shapes which is particular to topology can be described in terms of three points.

1. The properties of a shape studied by topology are independent of any particular coordinate representation of the shape in question, and instead depend only on the pairwise distances between the points making up the shape.
2. The topological properties of shapes are *deformation invariant*, i.e. they do not change if the shape is stretched or compressed. They would of course change if non-continuous transformations are applied that "tear" the space.
3. Topology constructs compressed representations of shapes, which retain many interesting and useful qualitative features while ignoring some fine detail.

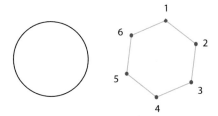

Thus, topology deals with shapes in two distinct ways. The first is by building compressed combinatorial representations of shapes, via processes such as triangulation. Of course some information about a shape is lost in this discretization, such as fine-scale curvature information, but, as in the example above, the rough overall structure is preserved in passing from the circle to the hexagon. The second way is by attempting to measure shapes, or aspects of shapes. This is done via *homological signatures*, which permit a kind of count of the occurrences of patterns within a shape. The adaptation of these signatures to the study of point cloud data (sets of data points in space) is the subject of this book.

A motivating example comes from contemplation of the phase space pictures of the Lotka–Volterra equations. Recall that these equations describe the population dynamics in a simple predator–prey model and result in oscillatory behavior, which gives rise to loops in phase space. One could of course describe such a loop by giving a precise parametrization (i.e. a system of local coordinates) for the loop. However, for many purposes the fact that the shape of the phase portrait is a smooth deformation of a circle is the most salient detail. More generally, consider the family of examples coming from the mapping of a circle to Euclidean space under a wide variety of embeddings. A salient qualitative description would extract the fact that the underlying data comprised a circle. The intuitive idea behind algebraic topology is that one should try to distinguish or perhaps even characterize spaces by the occurrences of such qualitative patterns within a space. For the Lotka–Volterra example one could say that a characteristic pattern is the presence of a loop in the space surrounding the empty region in the middle. One could say intuitively that the count of loops in the phase portrait is one, in that there is "essentially" only one loop in the space. The same characterization would hold for an annulus, where the essential loop winds around the central removed disc. It is not so easy to make mathematical sense of this observation, because there are often families of loops that we would regard as being essentially the same, as in the figure below.

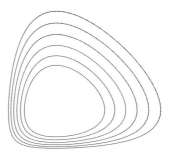

The presence of essentially one loop is something which a priori is difficult to quantify, since in fact there is an uncountable infinity of actual loops which have the same behavior, i.e. they each wind around the hole once. In order to resolve this difficulty and formalize the notion that there is essentially only one loop, we are forced to perform some abstract constructions involving equivalence relations to obtain a sensible way of counting the number of loops. The idea is that we must regard many different loops as equivalent, in order to get a count of the occurrences, not of each individual loop but, rather, of a whole class of equivalent loops. This step is responsible for much of the abstraction which has been introduced into the subject. Once that layer of abstraction has been built, it provides a way to detect the presence of geometric patterns of certain types. The general idea of a pattern is of course somewhat diffuse, with many different meanings in many different contexts. In the geometric context, we define patterns as maps from a template space, such as a circle, into the space. A large part of the subject concerns the process of reducing the abstract constructions described above to much more concrete mathematical constructions, involving row and column operations on matrices. The goals of the present volume are the following.

- To introduce the pattern detection signatures which come up in algebraic topology, and to simultaneously develop the matrix methods which make them into computable and usable invariants for various geometric problems, particularly in the domain of *point clouds* or *finite metric spaces*. We hope that the introduction of the relevant matrix algorithms will begin to bridge the gap between topology as practiced "by hand" and the computational world. We will describe the standard methods of homology, which attach a list of non-negative integers (called Betti numbers) to any topological space, and we also discuss the adaptation of homology to a tool for the study of point clouds. This adaptation is called *persistent homology*.

- To introduce the mathematics surrounding the collection of *persistence barcodes* or *persistence diagrams*, which are the values taken by persistent homology constructions. Unlike Betti numbers, which are integer valued, persistent homology takes its values in multisets of intervals on the real line. As such, persistence barcodes have a mix of continuous and discrete structure. The study of these spaces from various points of view, so as to be able to make them maximally useful in various problem domains, is one of the most important research directions within applied topology.
- To describe various examples of applications of topological methods to various problem domains.

1.1   Overview

The purpose of this book is to develop topological techniques for the study of the qualitative properties of geometric objects, particularly those objects which arise in real-world situations such as sets of experimental data, scanned images of various geometric objects, and arrays of points arising in engineering applications. The mathematical formalism called *algebraic topology*, and more specifically *homology theory*, turns out to be a useful tool in making precise various informal, intuitive, geometric notions such as holes, tunnels, voids, connected components, and cycles. This precision has been quite useful in mathematics proper, in situations where we are given geometric objects in closed form and where calculations are carried out by hand. In recent years, there has been a movement toward improving the formalism so that it becomes capable of dealing with geometric objects from real-world situations. This has meant that the formalism must be able to deal with geometric objects given via incomplete information (i.e. as a finite but large sample, perhaps with noise, from a geometric object) and that automatic techniques for computing the homology are needed. We refer to this extension of standard topological techniques as *computational topology*, and it is the subject of this volume.

We will assume that the reader is familiar with basic algebra, groups, and vector spaces.

In this introductory chapter, we will sketch all the main ideas of computational topology, without going into technical detail. The remaining chapters will then include a precise technical development of the ideas as well as some applications of the theory to actual situations.

1.2   Examples of Qualitative Properties in Applications

1.2.1   Diabetes Data and Clustering

Diabetes is a metabolic disorder which is characterized by elevated blood glucose levels. Its symptoms include excessive thirst and frequent urination. In order to understand the disease more precisely, it is important to understand the possible configurations of values that various metabolic variables can exhibit. The kind of understanding we hope for is geometric in nature. An analysis of this type was carried out in the 1970s by Reaven & Miller (1979).

   In this study, a collection of five parameters (four metabolic quantities and the relative weight) were measured for each patient. Each patient then corresponds to a single data point in five-dimensional space. In Reaven & Miller (1979), the *projection pursuit* method was used to produce a three-dimensional projection of the data set, which looks like the situation on the left in Figure 1.1.

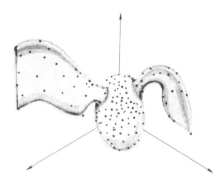

Figure 1.1 Diabetes patient distribution. See the main text for a description of the figure. From Reaven & Miller (1979), reproduced with permission of Springer–Nature, © 1979.

   Each patient was classified as being normal, chemical diabetic, or overt diabetic. This is a classification which physicians devised using their observation of the patients. It was observed that the normal patients occupied the central rounded object, and the chemical diabetics and overt diabetics corresponded to the two "ears" in the picture. Another very interesting method for visualizing the set was introduced in Diaconis & Friedman (1980). These visualizations suggest that the two forms of diabetes are actually fundamentally different ailments. In fact, physicians have understood that diabetes occurs in two forms, "Type I" and "Type II". Type I patients often have juvenile onset, and the disease may be independent of the patient's life style choices. Type II diabetes more often occurs later in life and appears to depend on life style choices. The chemical diabetics are likely to be thought of as Type II diabetics who eventually may arrive at the overt diabetic stage, while the overt diabetics might arrive at overt status directly.

   In this case, the qualitative property of the figure that is relevant is that it has the two distinct ears, coming out of a central core. Although human beings can recognize this fact in this projection, it is important to formalize mathematically what this means, so that one can hope to automate the recognition of this qualitative property. For example, there may be data sets for which no two- or three-dimensional projection gives a full picture of the nature of the set. In this case, the mathematical version of this statement would be as follows. We suppose that the three categories of patients (normal, chemical, and overt) correspond to three different regions A, B, and C in five-dimensional Euclidean space.

What this experiment suggests is that if we consider the union $X = A \cup B \cup C$ (which corresponds to all patients) and then remove A, the region corresponding to the normal patients, the region we are left with breaks up into two distinct connected pieces, which do not overlap and in fact are substantially removed from each other. Clustering techniques from statistics were used in Symons (1981) to find methods to differentiate between these two components. The qualitative question above about the nature of the disease can now be stated as asking how many connected components are present in the space of all patients having some form of diabetes. Finding the number of connected components of a geometric object is a topological question.

1.2.2   Periodic Motion

Imagine that we are tracking a moving object in space, and that the information is given in terms of a three-dimensional coordinate system, so that we are given coordinates $(x(t), y(t), z(t))$. If we want to know whether the object is moving periodically, say, because it is orbiting around a planet, we can simply check whether the values of the coordinates repeat after some fixed period of time. Suppose, however, that we are not given the time values corresponding to the points but just a set of positions, and want to determine whether the object is undergoing periodic motion. We would thus like to know whether the set of positions forms a closed loop in space. If the object is orbiting around a single planet or star, and we therefore know by Kepler's laws that the geometric shape of the orbit must be that of an ellipse, we can determine that the object is orbiting by simply curve-fitting an ellipse to the data set of positions. Suppose, however, that the object is being acted on gravitationally by several other objects, so that the path is not a familiar kind of closed curve. We would then still want to know whether the space of positions is a closed loop, but perhaps not one for which we have a familiar set of coordinatizations. The qualitative property in which we are interested is whether the space is a closed loop, and we would like to develop techniques which allow us to determine this without necessarily asking for a particular coordinatization of the curve. In other words, we are asking whether our space is a closed loop of some kind, not exactly what type of loop it is.

A more difficult situation is where we are not actually given the values of the position of the object but, rather, a family of images taken from a digital camera. In this case, the set of these images actually lies in a very high-dimensional space, namely the space of all p-vectors, where p is the number of pixels. Each pixel of each image is given a value, the gray-scale intensity at that pixel, and so each image corresponds to a vector, with a coordinate for each pixel. If we take many images sequentially, we will obtain a family of points in the p-dimensional space, which lies along a subset which should be identified topologically with the set of positions of the object, i.e. a circle. So, although this set is not identified with a circle through any simple set of equations in p variables, the qualitative information that it is a circle is contained in this data. This is an example of an exotic coordinatization of a space (namely the circle) and shows that, in order to analyze this kind of data, it would be very useful to have tools which can tell whether a space is a closed loop, without its having to be any particular loop. In other words, coordinate-free tools are very useful.

1.2.3   Curve and Shape Recognition

There are situations where we have geometric objects which do not come from experimental data, but where qualitative and coordinate free tools are of value for their analysis. Consider the problem of recognizing hand-printed characters. Hand-printed versions of a particular letter or number can vary a great deal. In fact, there exists a database (the MNIST database Bottou et al. 1994) which comprises many different handwritten versions of the numerals from 0 to 9. The variability comes from the fact that different people develop slightly different versions of the same character, and in fact these versions are sufficiently different that they may sometimes be used to identify the person who wrote them. Differences may also arise from the fact that one may not be looking directly, i.e. head-on, at the paper where the character is drawn, or that it may not be drawn on a flat surface. However, there are a sufficient number of qualitative cues which allow human beings to identify characters despite this variability. For example, if we compare the letter "A" with the letter "B", it is not hard to see that the letter "A" has a single closed loop in it, while "B" has two. Thus, the number of loops is a sufficient criterion to distinguish between these two letters, and it suggests the potential value of developing rigorous and automatic methods for determining the number of loops. Suppose instead that we consider the problem of distinguishing between the letter "U" and the letter "V". In this case, neither letter has a loop, but "V" has a "corner" and "U" does not. This is another useful qualitative cue. Finally, if we attempt to distinguish "C" from "I", we see that neither letter has a loop, and further that there are no corners, but that "C" has a curved arc while "I" does not. This is again a useful qualitative cue, which it will be useful to formalize.

Similar cues can allow us to distinguish between two-dimensional objects in $\mathbb{R}^3$, i.e. to perform *shape recognition*. For example, to distinguish the sphere from the torus (the two-dimensional surface of a doughnut), we can observe that every loop on the sphere can be contracted down to a point, while the torus has two obvious types of loop which cannot.

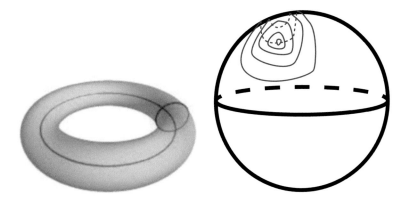

Similarly, one can distinguish between a tetrahedron and a cube by noting that a cube has eight vertices and 12 edges, while the tetrahedron has four vertices and six edges. Note that both these criteria are robust in the sense that if we make smooth deformations of the objects in question, these characteristics still remain unchanged.

We will see later that cues involving "corners", "curved arcs", "vertices", and "edges" are not directly topological. We will develop methods for recognizing these cues topologically on new spaces that we have constructed from the old ones using tangential information.

2 Data

In this chapter, we will discuss the properties and methods associated with various important data types, in order to indicate the wide range of applications for all methods of data science. Since the purpose of this volume is to leverage geometric structures which are applicable to data, we will point out the relevant geometric structures in each case. We will also discuss the conventional methods that are applicable in each situation. In particular, we will see that in many of these situations, it is important to use structures on the set of features attached to data sets, and we will point them out. In particular, geometric structures are often relevant for the features as well as for the data points themselves. We are certainly not claiming to be exhaustive in what we present; rather, we are attempting to give a reasonable sample of what is possible.

2.1 Data Matrices and Spreadsheets

Perhaps the most common representation of data sets is as tidy data: a spreadsheet with real numerical entries. Each data point corresponds to a row in the spreadsheet, and each column corresponds to a feature in the data set. In more mathematical terms, the data set is represented as a matrix of real numbers, where the number of rows is the number of data points and the number of columns is the number of features in the spreadsheet. This interpretation suggests that linear algebra should be useful in the analysis of data sets, and this is indeed the case.

An important method for the study of data sets given in matrix form is *principal component analysis* (PCA); see Hastie et al. (2009) for a comprehensive description. It is a method for modeling data by finite (but possibly very large) subsets of inner product spaces. Of course, if we are given a data matrix, the data is exactly identified with a finite subset of an inner product space, namely $V = \mathbb{R}^N$, where N is the number of columns of the matrix. The inner product on V is the standard inner product, in which the unit basis vectors form an orthonormal basis. If N is small, PCA permits some useful kinds of analysis. If $N = 1, 2$, or 3, then one can actually visualize the data set as a scatter plot. Even if N is greater than 3, but is of relatively small size, PCA makes many kinds of calculations possible and can give insight. This is the case in the text analysis of corpora of documents, where the data sets might have nearly a million features (one for every word in a dictionary), but where one can often reduce the model to a few hundred features, many of which are interpretable. In summary, PCA enables visualization and dimensionality reduction.

Let $X = \{\vec{v}_1, \ldots, \vec{v}_M\} \subseteq V$ be a finite subset in an N-dimensional inner product space V. We may assume that the vector of means $\hat{\vec{v}} = \frac{1}{M} \cdot \sum_i \vec{v}_i = 0$ for, if it is not, we may replace X by the subset $X^0 = \{\vec{v}_1 - \hat{\vec{v}}, \vec{v}_2 - \hat{\vec{v}}, \ldots, \vec{v}_M - \hat{\vec{v}}\}$ and note that X may be obtained from X^0 by simply adding the vector $\hat{\vec{v}}$ elementwise. We will assume that $\hat{\vec{v}} = 0$. Given a set $X \subseteq V$, where V is an inner product space, the *variance* of X is the mean $\frac{1}{|X|} \sum_{\vec{v} \in X} \vec{v} \cdot \vec{v}$. Given any subspace $W \subseteq V$, the inner product yields a direct sum decomposition $V = W \oplus W^\perp$, and therefore X produces a subset $X_W \subseteq W$ by projection onto the factor W. This means that each subspace W gives a lower-dimensional model for the data set X. If we fix a dimension $n \leq N$, the question is how to choose the best n-dimensional model W. The heuristic used in PCA is that the most informative choice of W is the one which maximizes the variance of X_W. It turns out that for each $0 < n \leq N$ this optimization problem always has a solution, although the solution may not always be unique. Furthermore, one can solve the optimization problem efficiently, thereby creating a family of models, one for each dimension. The solution method goes through the so-called *singular value decomposition* (SVD) of a matrix P, which asserts that, for any $M \times N$ matrix P, there are orthogonal matrices U and V, where U is $M \times M$ and V is $N \times N$, such that the matrix UPV^T is of the form

$$\left[\begin{array}{c|c|c} \Lambda_r & 0 & 0 \\ \hline 0 & 0 & 0 \end{array}\right],$$

where Λ_r is an $r \times r$ diagonal matrix of the form

$$\begin{bmatrix} \lambda_1 & 0 & 0 & 0 \\ 0 & \lambda_2 & 0 & 0 \\ 0 & 0 & \ddots & \vdots \\ 0 & 0 & \cdots & \lambda_r \end{bmatrix},$$

with $\lambda_1 \geq \lambda_2 \geq \cdots \geq \lambda_r \geq 0$. The λ_i are called the *singular values* of P. The indeterminacy in the choice of the decomposition referred to above is due to the possibility of equalities among the singular values. Computationally, the SVD can be computed using eigenvalue techniques, since the matrices U and V may be identified with eigenbases for MM^T and $M^T M$, respectively. See Yanai et al. (2011) for a thorough treatment.

Principal component analysis is a method for the unsupervised analysis of data. In many cases it is desirable to perform a more supervised analysis, for example to estimate a collection of outcome variables using a set of independent variables, where the outcome variables and independent variables are disjoint. Standard *linear regression* is a supervised method which performs this optimization. These methods are "workhorses" for many applied mathematical and statistical problems. They are part of a large area of study within applied mathematics and statistics, and the reader may consult Montgomery et al. (2006), for example, to see what is possible with them.

Support vector machines provide another supervised method, whose input is a collection X of points in a Euclidean inner product space V, together with a partition of

X into two classes, X_{+1} and X_{-1}, and whose goal is to produce a simple mathematical procedure (called a *linear classifier*) which predicts the subset to which a new data point should belong. The formula is produced by considering the decomposition of V using a hyperplane, where one class lies on one side of the hyperplane and the other on the other side. To sketch the idea, we suppose first that the sets X_{+1} and X_{-1} are in fact *linearly separable*, i.e. that there is a hyperplane such that X_{+1} and X_{-1} are entirely on opposite sides of it, with no point of X actually lying on it. We define a set $\mathfrak{H}$ to be the collection of all pairs of parallel hyperplanes (H_{+1}, H_{-1}) such that X_{+1} lies on the opposite side of H_{+1} from H_{-1}, and such that X_{-1} lies on the opposite side of H_{-1} from H_{+1}. Each such pair of hyperplanes determines a classifier, as follows. We consider the plane H_0 which is "halfway between" H_{+1} and H_{-1}, and describe it by an equation of the form $\varphi(\vec{v}) = \vec{w} \cdot \vec{v} - b = 0$. The parallel planes H_{+1} and H_{-1} are then given by equations $\vec{w} \cdot \vec{v} - b_+ = 0$ and $\vec{w} \cdot \vec{v} - b_- = 0$ and, after possibly multiplying the equation by -1, we may assume that $b_+ > b > b_-$ and that $b_+ - b = b - b_-$. The classification is now achieved by asserting that a vector $\vec{v}$ belongs to (a) X_{+1} if $\vec{w} \cdot \vec{v} - b > 0$ and (b) X_{-1} if $\vec{w} \cdot \vec{v} - b < 0$. Of course, this can works for many pairs of hyperplanes. However, for each pair of parallel hyperplanes $(H_+, H_-) \in \mathfrak{H}$, there is a well-defined *distance* between the hyperplanes, and the choice that is made for the classifier is to select the pair $(H_+, H_-) \in \mathfrak{H}$ which maximizes this distance (it is unique), and then to use the classifier attached to this pair. This would arguably be the best way to choose a linear classifier for new data points for which a predicted class is desired.

Not all $X = X_{+1} \cup X_{-1}$ are linearly separable, in which case the above construction is not applicable. However, it gives the motivation for a different optimization problem which operates in non-separable situations. The idea is to define a loss function attached to a hyperplane and a set of points divided into two classes within a finite-dimensional inner product space. We note first that we will be optimizing over pairs $(\vec{w}, b)$, where $\vec{w} \in V$ and b is a real number. This data represents the hyperplane $\vec{w} \cdot \vec{v} - b = 0$. For each data point $x_i \in X$, we assign $y_i \in \{\pm 1\}$ by declaring that $y_i = 1$ if $x_i \in X_{+1}$ and $y_i = -1$ if $x_i \in X_{-1}$. To this configuration we now associate a loss function

$$\mathfrak{L} = (i, \vec{w}, b) = \max(0, 1 - y_i(\vec{w} \cdot x_i - b))$$

with each data point x_i. We note that $\mathfrak{L} = 0$ if either $y_i = +1$ and $\vec{w} \cdot x_i - b \geq 1$ or $y_i = -1$ and $\vec{w} \cdot x_i - b \leq -1$. This means that the loss function is zero for points $x_i \in X_{+1}$ which are correctly correctly classified by the classifier $\vec{w} \cdot x_i - b \geq 1$, and similarly for points in X_{-1}. Points which are not classified by either of these classifiers lie between the hyperplanes given by $\vec{w} \cdot \vec{v} - b = 1$ and $\vec{w} \cdot \vec{v} - b = -1$ and are assigned a positive loss value depending on how close they are to one or the other hyperplane. The *hard loss function* for the entire configuration is given by the sum

$$\sum_{i=1}^{N} \mathfrak{L}(i, \vec{w}, b),$$

which one can attempt to minimize. If the subsets X_{+1} and X_{-1} are linearly separable then the value 0 can be achieved, as one can easily see from the linearly separable analysis above. One could now decide simply to use this loss function even in the linearly inseparable case, but it turns out to be useful to introduce a parameter λ and consider instead the modified loss function

$$\frac{1}{N}\left[\sum_{i=1}^{N}\mathfrak{L}(i,\vec{w},b)\right]+\lambda\|\vec{w}\|.$$

The aim here is that there should be a penalty for $\|\vec{w}\|$ being too large. The reasoning is as follows. A hyperplane can be represented by many equations of the form $\vec{w}\cdot\vec{v}-b=0$, since we may multiply the equation by any non-zero real number C and obtain an equally valid equation. What does change with C are the hyperplanes $\vec{w}\cdot\vec{v}-b=\pm1$, which become closer to each other as $C\to\infty$. In the infinite limit, this means that the "band" between them shrinks until the two hyperplanes become one. Thus the hard loss function is essentially just an evaluation of the fraction of points that are misclassified. This is in general too rigid, since one wants the loss function not to change too much with small changes in the positions of the data points. The parameter λ allows one to tune the classifier, to arrive at a model which is more robust to such small changes.

This method can be extended to classification problems with more than two classes. In addition, one might hope to extend it to non-linear situations. This is sometimes done by embedding the problem non-linearly into a much higher-dimensional inner product space and performing the linear classification there. See Vapnik (1998) for a more detailed discussion.

Another method for addressing classification problems with linear algebra is *logistic regression*. In the simplest case, this method requires a collection of data points which consist of a number of continuous variables, called the independent variables, and one $\{0,1\}$-valued outcome variable. The aim is to design a procedure that estimates the probability that the outcome variable equals 1, given the values of the independent variables. It is assumed that the probability has the form $\sigma(\sum_i c_i x_i + b)$, where the x_i are the independent variables, the c_i and b are parameters to be estimated, and σ is the logistic function, given for a real number t by

$$\sigma(t)=\frac{1}{1+e^{-t}}.$$

In order to fit a model, we must choose a measure of the model fit that we can optimize. The standard choice is the *maximum likelihood* function. We suppose that the data points are $\{\vec{x}_1,\ldots,\vec{x}_N\}$, and that the outcome for the point $\vec{x}_i$ is y_i such that $y_i\in\{0,1\}$. Given the model determined by the coefficient vector $\vec{c}$ and the number b, the likelihood that the outcome variable equals 1 at the data point $\vec{x}_i$ is given by $h_{\vec{c},b}(\vec{x}_i)=\sigma(\vec{c}\cdot\vec{x}_i+b)$ and the likelihood that it equals 0 is $1-h_{\vec{c},b}(\vec{x}_i)$. The likelihood of the values being correct for all values of i is therefore

$$\prod_{i\,|\,y_i=0}(1-h_{\vec{c},b}(\vec{x}_i))\prod_{i\,|\,y_i=1}h_{\vec{c},b}(\vec{x}_i),$$

or, equivalently,

$$\prod_i(1-h_{\vec{c},b}(\vec{x}_i))^{(1-y_i)}h_{\vec{c},b}(\vec{x}_i)^{y_i}.$$

This function is now maximized using gradient descent methods. Logistic regression can also be extended to classification problems with more than two classes. A good discussion can be found in Hastie et al. (2009).

2.2   Dissimilarity Matrices and Metrics

There is a straightforward extension of the notion of distance in $\mathbb{R}^2$ and in $\mathbb{R}^3$ to $\mathbb{R}^n$, via the formula

$$d(\vec{v}, \vec{w}) = \sqrt{(\vec{v} - \vec{w}) \cdot (\vec{v} - \vec{w})}.$$

Therefore, when we have a data set of points $S = \{\vec{v}_1, \ldots, \vec{v}_N\} \in \mathbb{R}^n$, we create a symmetric matrix of distances, $\mathfrak{D}(S)$, given by

$$\begin{bmatrix} d(\vec{v}_1, \vec{v}_1) & \cdots & d(\vec{v}_1, \vec{v}_N) \\ \vdots & & \vdots \\ d(\vec{v}_N, \vec{v}_1) & \cdots & d(\vec{v}_N, \vec{v}_N). \end{bmatrix}$$

This matrix can be thought of as a *dissimilarity matrix*, because large distances can be thought of as indicating dissimilarity and small distances as indicating similarity.

Definition 2.1   A *dissimilarity matrix* is a non-negative symmetric matrix which has zeros along the diagonal. It is also useful to think of it as a structure on a finite set X, defining a function $\mathfrak{D} \colon X \times X \to [0, +\infty)$ which (a) vanishes on the diagonal and (b) has the symmetry property $\mathfrak{D}(x, x') = \mathfrak{D}(x', x)$. A *dissimilarity space* is a pair $(X, \mathfrak{D})$, where X is a set and $\mathfrak{D}$ is a non-negative real-valued function on $X \times X$ satisfying conditions (a) and (b) above. A dissimilarity matrix is *of metric type* if additionally (a) all its off-diagonal entries are non-zero and (b) it satisfies the triangle inequality

$$\mathfrak{D}_{ik} \leq \mathfrak{D}_{ij} + \mathfrak{D}_{jk}.$$

The corresponding functions on the set $X \times X$, where X is the set of columns of the matrix $\mathfrak{D}$, is then called a metric, and the pair consisting of X and the function $\mathfrak{D}$ is called a metric space.

There will be situations where the data we are working with are not necessarily embedded in Euclidean space, but where we can nevertheless construct a dissimilarity matrix. Here are some examples of when this occurs.

1. There are a number of abstract distance functions that can be defined on various kinds of categorical data, such as Hamming distances, which do not obviously arise from subsets of Euclidean space.
2. If we have points that we think of as sampled from a Riemannian manifold M embedded in Euclidean space, their dissimilarities may be better modeled by geodesic distances in M rather than the Euclidean distances. For data embedded in $\mathbb{R}^n$, there are versions of the geodesic distance which are closely related to the graph distances between data points, where points sufficiently close are connected to give

a graph structure. This distinction is the key to the utility of the ISOMAP algorithm (Tenenbaum et al. 2000).

3. There are situations where human beings have constructed intuitive but quantitative notions of similarity, and it is desirable to work with them geometrically. See for example Sneath & Sokal (1973) for numerous biological examples.

It is therefore useful to create methods and algorithms which operate directly on dissimilarity matrices and which do not require a vector representation.

The first such algorithm that we will discuss is *multidimensional scaling* (MDS) (Borg & Groenen 1997). Given an $n \times n$ dissimilarity matrix, interpreted as a dissimilarity measure on a set of points $S = \{x_1, \ldots, x_2\}$, MDS produces an embedding of S in Euclidean space $\mathbb{E}^d$, for some d, such that the loss function

$$\sum_{i<j} (\|\tilde{x}_i - \tilde{x}_j\| - d(x_i, x_j))^2$$

is minimized, where $\tilde{x}_i$ denotes the vector in $\mathbb{E}^d$ corresponding to $x_i \in S$. In other words, the dissimilarity matrix is approximated as well as possible by the distances between the embedded points in Euclidean space. It is possible to choose other loss functions, depending on the situation, but this one has the attractive property that it is predictable in the sense that it produces a predictable minimum in the dissimilarity matrices which come from Euclidean space. This is demonstrated by the following analysis. Let $\mathfrak{D}$ be a matrix of Euclidean distances, $d_{ij} = \|\tilde{x}_i - \tilde{x}_j\|$. We form the associated matrix K:

$$K = -\frac{1}{2}HDH,$$

where $H = I - \frac{1}{n}\mathbf{1}\mathbf{1}^T$ and $\mathbf{1}$ denotes the vector of all ones). When $\mathfrak{D}$ is a Euclidean distance matrix, K is positive semi-definite and so standard minimization arguments imply that the solution is given by $\Lambda^{1/2}Z^T$, where Λ is diagonal with the top k eigenvalues of K along the diagonal and Z contains the corresponding top k eigenvectors of K. This also demonstrates that the result is the same as that obtained by using PCA on the Euclidean data. Multidimensional scaling can be used for dimensionality reduction by choosing the dimension of the Euclidean space into which to embed. However, a significant advantage of MDS is the fact that the dissimilarity matrix used as input need not come from a distance function but can simply arise from any symmetric dissimilarity matrix (Borg & Groenen 1997). It is apparent that MDS is an unsupervised method for data analysis, and the discussion above shows that it generalizes PCA.

Another class of methods for unsupervised data analysis is *cluster analysis*. In this case the output of the method is a partition of the data set rather than an embedding in a Euclidean space. There is a very rich family of methods that perform this task, and we will not attempt to be comprehensive here but, rather, will give a few simple examples. For a more complete treatment, see Everitt et al. (2011).

By a *clustering algorithm*, we mean an algorithm which takes as its input a dissimilarity space $(X, \mathfrak{D})$, and from it defines a partition of the set X. The blocks of the partition are called *clusters*. Here are some examples.

Example 2.2   *Single-linkage clustering with scale parameter R* is defined on a dissimilarity space $(X, \mathfrak{D})$ by declaring that two points x and x' are in the same cluster if and only if there is a sequence $x_0, x_1, \ldots, x_n$ of elements of X with the properties that (a) $x_0 = x$, (b) $x_n = x'$, and (c) $\mathfrak{D}(x_i, x_{i+1}) \leq r$ for all i. This defines a partition of X, as is readily verified. The method is conceptually very simple but requires a choice of r, for which there may be no particular conceptual justification. We will return to this point shortly.

Example 2.3   *K-means clustering* is restricted to subsets of Euclidean space, and requires that one initially selects a number (K) of clusters. It proceeds by optimizing a particular objective function F, which is defined as follows. Suppose that we are given a family of K clusters S_i, and let μ_i denote the mean of the vectors in S_i. We then define $F(\{S_i\}_i)$ to be the sum

$$\sum_{i=1}^{K} \sum_{v \in S_i} \|v - \mu_i\|^2.$$

Example 2.4   Single-linkage clustering is dependent on the choice of the scale variable r. It is difficult to decide how to choose this threshold. Statisticians have shown that there is a mathematical structure which can be computed effectively and which gives information about clustering at all values of r at once. It is called a *dendrogram*, and a picture of one is given in Figure 2.1.

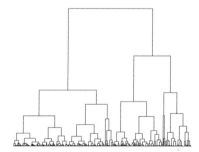

Figure 2.1 Dendrogram for hierarchical clustering; the figure shows the average linkage on the Palmer Penguins data set (Horst et al. 2020).

Dendrograms will be defined precisely in Section 4.1, but they have the property that if we take the cross-section of a horizontal line at height r, as in Figure 2.1, the collection of points in the intersection of the horizontal line with the dendrogram is in one-to-one correspondence with the collection of clusters obtained at threshold level r. Moreover, we can observed how the clusters merge as r increases. This construction is called *single-linkage hierarchical clustering*.

Example 2.5   There is another way to think about single-linkage hierarchical clustering, which constructs the dendrogram iteratively using the concept of merging clusters. Given a pair of clusters, merging them means creating a single new cluster which is the union of the two clusters. For a dissimilarity space $(X, \mathfrak{D})$, the first step is to merge pairs of points whose distance is the minimum positive value of $\mathfrak{D}$. At this point, we have a clustering of X. Each cluster is a subset of X and, for each pair (ξ, ξ') of clusters, we can evaluate

$$\mathfrak{L}(\xi, \xi') = \min_{x \in \xi, x' \in \xi'} \mathfrak{D}(x, x').$$

We can now merge all pairs of clusters where the minimum positive value of $\mathfrak{L}$ is achieved, and this process can be iterated to yield a family of clusterings of increasing coarseness. By keeping track of the values $\mathfrak{L}$ computed at each stage, one can recover single-linkage hierarchical clustering in this way. This point of view permits a useful generalization, on using other choices for the function $\mathfrak{L}$. For example, one can replace $\mathfrak{L}$ by $\mathfrak{L}^{\text{ave}}$, defined by setting $\mathfrak{L}^{\text{ave}}(\xi, \xi')$ equal to the average of the set of values $\{\mathfrak{D}(x, x')\}$ over pairs $x \in \xi$ and $x' \in \xi'$, where $\xi \neq \xi'$. One can also replace $\mathfrak{L}$ by the function $\mathfrak{L}^{\text{max}}$, defined by setting $\mathfrak{L}^{\text{max}}(\xi, \xi')$ equal to the maximum of the set of values $\{\mathfrak{D}(x, x')\}$ over pairs $x \in \xi$ and $x' \in \xi'$, for any pairs of distinct clusters (ξ, ξ'). Each choice of $\mathfrak{L}$ gives a different hierarchical clustering scheme, with $\mathfrak{L}$, $\mathfrak{L}^{\text{ave}}$, and $\mathfrak{L}^{\text{max}}$ corresponding to hierarchical schemes referred to as *single-linkage*, *average-linkage*, and *complete-linkage* hierarchical clustering respectively. Average-linkage and complete-linkage are often useful since they address problems arising with single-linkage problems, one being the *chaining problem*. This refers to the fact that single linkage often produces long sequences of clusters which consist of very small clusters adjoining one large one. See Everitt et al. (2011) for a discussion of this problem.

Cluster analysis is a highly developed area in data science, and we refer the reader to Everitt et al. (2011) for a thorough treatment.

2.3   Categorical Data and Sequences

Many data sets are not in numerical matrix format, and also they may not be naturally equipped with dissimilarity measures. Finding methods for putting them into matrix or dissimilarity format is of great value. Consider the following simple example. Suppose that we are given a data set whose entries are "shopping baskets" for a store. We are interested in understanding the set of these baskets in some way. Each entry in the data set is a list of products carried by the store, perhaps with multiplicities for purchases of multiple units, but the order in the entries is not meaningful for us because the order in which the items are rung up is not of significance. The data could be encoded as list of product identifiers. These identifiers might be numeric or alphanumeric, but a numerical quantity carries no meaning beyond the purpose of identification. As it stands, the methods in Sections 2.1 and 2.2 are of no use, because we do not have a numerical matrix representation, either as vectors or as a dissimilarity measure. Moreover, the data is given to us in a form which introduces extraneous information, namely the ordering of the items in the basket. We introduce *one-hot encoding* as a method for overcoming both of these problems. Let S be a finite set, and let X be a data set whose elements are subsets of S. We construct the vector space $\mathbb{R}^{\#(S)}$, with its standard basis $\{e_s\}_{s \in S}$, and assign to each subset $S_0 \subseteq S$ the vector $v(S_0) = \sum_s c_s e_s$, where $c_s = 1$ if $s \in S_0$ and $c_s = 0$ if

$s \notin S_0$. We have thus assigned a vector to each subset of S and consequently we have arrived at a data matrix, to which the matrix methods given above can be applied. The collection of subsets of S can now be viewed as a dissimilarity space, in which distances between subsets of S are determined by their symmetric differences. We note that the construction extends to multisets of elements, where elements can occur with multiplicities greater than one (see Hein 2003, for a discussion of these), in a straightforward way.

Let us consider another situation, where a company is studying its sales representatives (we will call them reps), and the reps have collected data consisting of pairs (rep, productid), where the two variables take values in the discrete sets of reps and company products, respectively. Each data point represents one transaction generated by the rep. The company would like to have information about the performance of its reps. The given list is not informative about reps directly, since each data point corresponds to only a single transaction. We would instead like to create a data set where each data point corresponds only to a rep, and carries information about the distribution of products the rep has sold. A method for doing this is via *pivot tables*.

For each rep, we have a multiset of products. Let us say the ith rep is associated with a set S_i of products. The *pivot table transform* associated with a variable salesrep begins with a data set of transactions which are of the form

$$(\text{rep}, p),$$

where p is a product. Note that a particular value rep can occur in many data points, one for each transaction. The pivot table transform now associates with the data set of transactions a data set where the entries are of the form

$$(\text{rep}, \{p_1, p_2, \ldots, p_n\})$$

and where each value rep of salesrep can occur only once and the second coordinate, within braces, is a multiset of products rather than a single product. The second coordinate is the multiset of all products that have been sold by the sales representative rep. To produce a matrix representation, we can apply one-hot encoding and obtain a data set consisting of elements (rep, v_{rep}). It corresponds to an $m \times n$ matrix, where m is the number of sales reps and n is the number of all possible products. This transform is extremely useful, and can also be used in situations where some variables are actually numerical instead of belonging to a discrete set. Suppose, for example, that our data set of transactions instead consisted of pairs (rep, x), where x is the price of the item in a transaction. Then we could form the pivot table transform associated with the variable salesrep, and obtain a pair

$$(\text{rep}, \{x_1, \ldots, x_n\}),$$

where $\{x_1, \ldots, x_n\}$ is the set of all prices for all transactions involving the sales representative rep. Since we might be interested in the total value of the transactions involving rep, we might here apply a different method to obtain a vector, in this case assigning to the set $\{x_1, \ldots x_n\}$ the sum $x_1 + \cdots + x_n$. If we were interested in the average value of the transactions for a given rep, we would instead associate the average value of the x_i. One could also consider the maximum or minimum values of the x_i.

The pivot table transform is in general defined in two steps. We suppose that we are given a data set X consisting of coordinates x_s, where $s \in S$ for some finite set S. The quantities s may correspond to variables of different types, namely one might consist of elements of a discrete set, another might consist of real numbers, another of positive integers, yet another of subsets of a fixed discrete set, etc. Suppose that S is the disjoint union of two sets S_0 and S_1. An S_i-*vector* is a tuple $\{x_s\}_{s \in S_i}$, where x_s has the type of the variable s. For each data point $x \in X$, we have two projections $\pi_0(x)$ and $\pi_1(x)$, where $\pi_i(x)$ is an S_i-vector for $i = 0, 1$. The first step in the pivot table transform associates with X a set $\mathfrak{P}(X)$ consisting of pairs (ξ, S), where ξ is an S_0-vector and S is a set of S_1-vectors. The set $\mathfrak{P}(X)$ has exactly one element for each unique value taken by $\pi_0(x)$ as x ranges over X, and, for a fixed ξ, the multiset S consists of all the S_1-vectors that occur as $\pi_1(x)$ for data points $x \in X$ for which $\pi_0(x) = \xi$. The second step is the *vectorization step*, which applies a method for assigning vectors in a vector space to sets of S_1-vectors. The use of one-hot encoding above, as well as the sum, average, max and min methods for sets of real numbers, are all such methods but one can imagine many others.

In the examples above, we have examined situations where we have collections and are not interested in the ordering of the elements in the collection. There are many situations where one *is* interested in the ordering. A notable example is in genomics, where one considers sequences of elements belonging to an alphabet, such as the 20-element alphabet of amino acids. In this case, there is a simple vectorization step. Suppose that we are considering sequences of length N in an alphabet A of cardinality a. The we may use one-hot encoding to assign a vector $h(a)$ of length a to members a of the alphabet A. We may then assign to the sequence $(a_1, \ldots a_N)$ the vector $(h(a_1), \ldots h(a_N))$, which is a vector of length aN. This is a simple vectorization method, but in this case it is often more useful to construct dissimilarity measures directly. For two sequences σ and σ' of length N in the alphabet A, we define the *Hamming distance* between σ and σ' to be the number of elements $i \in \{1, \ldots, N\}$ such that $\sigma_i \neq \sigma'_i$. This is clearly a dissimilarity measure, and it is in fact of metric type. This dissimilarity measure, and variants of it, are very useful in many situations, particularly in genomics. See Gusfield (1997) for a comprehensive treatment.

2.4   Text

A very interesting data type arises in natural language processing, namely that of a *corpus of documents*. Each document is a sequence of words, and they may be of varying length. It is very useful in for example in analyzing sentiment, or detecting trends in collection of newspaper articles. The general setup is that a corpus is a collection of documents and each document is a sequence of words, perhaps with some punctuation symbols included. Since the documents could be regarded as sets of words (losing the information about their order in the sequence), it is possible to use one-hot encoding directly, where a basis element in a vector space would be associated with each word. What this effectively does is to assign to each document the collection of word counts within the document for every word within a dictionary. This is certainly a possible vectorization for the data set, but it is not satisfactory because very common words, such as the word "the", will dominate the occurrences of other meaningful words in the document. A solution to this problem is via *tf–idf* (term-frequency–inverse-document-frequency).

Let D be a collection of documents d, referred to as the *corpus*. In tf-idf methods, we assign to each word–document pair (w, d) the number of occurrences of the word w in the document d, and denote it by $\text{tf}(w, d)$, the *term frequency* of w in d. In order to deal with the problem of frequently occurring words, we will want to weight this count, and the idea is to do that on the basis of the number of documents in D that include w. The most standard choice is to multiply this count by the *inverse document frequency*, which is defined to be

$$\text{idf}(w, d) = \log \frac{N}{|\{d \in D \text{ such that } w \in d\}|}.$$

Assigning to each document d the vector $\{\text{tf}(w, d)\,\text{idf}(w, D)\}_w$ gives a data matrix with $\mathcal{D}$ rows and W columns, where $\mathcal{D}$ is the number of documents in the corpus and W is the number of words in the dictionary. Note that, with this vectorization, any word which occurs in every document will have an identically zero value, so that every column in the data matrix corresponding to such words will be identically zero and can therefore be ignored. The tf–idf approach has worked well in many situations; principal component analysis has been applied extensively to these data matrices with interesting results. Nevertheless, because the tf-idf approach treats each document as just a multiset of words, it does not take full advantage of all the structure that is present. One could, for example, begin to take advantage of the additional structure of the ordering by dealing with k-grams, i.e. sequences of k consecutive words.

Another direction in natural language processing is the use of *word embeddings*. A word embedding consists of an inner product space with an assignment w in the dictionary. A canonical choice given above would be one-hot encoding, where the vector space V is $\mathbb{R}^W$ and where the vector v_w would be the standard basis vector e_w in the vector space $\mathbb{R}^W$. The point of word embeddings is to embed the words in a vector space of much smaller dimension, so that the distances between the associated vectors in some way reflect correlations between words. It is possible to use idf reweighting in conjunction with any such word embedding. One standard method is to apply PCA and choose the top few hundred coordinates. Other popular examples include *word2vec* and *Glove*. Useful references for natural language processing are Manning & Schütze (1999) and Eisenstein (2018).

2.5 Graph Data

There is a large class of data where either (a) the data is equipped with a graph structure or (b) the data set consists of elements which are graphs. The notion of a graph has several variants. Graphs may be directed or undirected, and they may carry weights or labels attached to edges and/or vertices. We present a number of different examples to show what is possible.

1. The internet is an example of a data set W of type (a) above, where the graph is directed and whose vertices are web pages and edges are hyperlinks. The web pages are considered as a data set, and the presence of a graph structure on W allows for feature generation. For example, *page rank* and the *hubs and authorities construction* (see Easley & Kleinberg 2011) create numerical features on the vertex set. One can also use simple graph-theoretic properties to create additional numerical features, such as total degree, inbound degree, and outbound degree. One can also create graph valued features by constructing a neighborhood of radius k, or other graph-valued constructions of a local nature. Using this idea, one can view the collection W as a data set of type (b) from above. It is also possible to assign weights to edges and vertices on the basis of the traffic at a webpage or across a particular hyperlink.

2. Any molecule can be regarded as an undirected graph whose vertices are atoms and whose edges are bonds. One can also attach weights or labels to the vertices using atomic weight, atomic number, or element label. Databases of molecules are therefore an example of type (b) above. Of course, simple measures, such as the number of atoms or various aggregate quantities concerning the bonds, are available but they do not reflect the full content of the data. In addition, there are standard collections of features called SMILES (see Weininger 1988). We will discuss how to apply topological methods to generate new useful features systematically in Section 6.2.

3. Crystals are regularly arranged collections of atoms or molecules. The theory of crystals is a highly developed area within physics, chemistry, and mathematics. The high degree of regularity makes a sophisticated theory, involving complex symmetry groups, possible. However, many materials of interest are *amorphous*, in that the atoms are not arranged with precise regularity but nevertheless exhibit geometric structure, which has been studied in various ways. One can encode information about an amorphous solid via a graph structure where the vertices correspond to the constitutent atoms or molecules and where the edges are assigned on the basis of chemical or physical properties. Often one can actually recognize bonds between the atoms and assign edges to them. In other situations, one assigns edges on the basis of distance thresholds. In Section 6.10, we will see how such methods can be used to generate features attached to various amorphous solid classes and to demonstrate that they carry chemical and physical information.

4. In genomics, the concept of a phylogenetic tree is a very important one. It describes the pattern of evolution within various classes of organisms and is certainly of a great deal of value in the study of viruses. The notion of *moduli spaces of trees* is a useful one, and it is studied in detail in Rabadan & Blumberg (2019).

2.6   Images

Images present another very interesting data type. Here the data typically consists of rectangular arrays of pixels, with each pixel assigned either a gray-scale value or a collection of color intensities using one of a number of possible ways of encoding color. This means that image data sets can be expressed as matrices with MN columns, where the image consists of an $M \times N$ rectangular array of pixels. What we note, though, is that the matrix is equipped with a kind of geometric structure, whether as a grid graph where every node is connected to its four nearest neighbors or with a distance function as a subset of the plane. What this means is that not only do we have the matrix entries, but we have knowledge about which nodes are close to a given node. Moreover, the regular grid structure supports translations in the image. Methods for working with images use the grid structure in a number of ways, as follows.

1. It is often useful to smooth images by replacing a pixel value at p by a sum of other pixel values, weighted by the inverse of their distance from p.

2. *Convolutional neural networks* (CNNs) (see Aggarwal 2018) constitute a computational method for computing with image data sets. They use the grid structure to devise a relatively sparse neural network architecture that performs strikingly well for a number of image analysis tasks, particularly for classification.

3. CNNs also use the translation property to insure that the object recognition methods that they construct have the property that an object (say a cat) in the lower left-hand corner of an image is recognized in exactly the same way as a cat in the upper right-hand corner.

4. The grid structure also allows for feature generation, one example being the so-called *Gabor filters* (see Feichtinger & Strohmer 1998).

2.7   Time Series

Time series are data sets that consist of sequences of observations of some kind, typically made at regular intervals. The most common version involves real-valued observations, but vector-valued time series are also of great interest. One can also think of time series with observations in other data types, such as regularly occurring text documents. Just as images are data where the features are arranged geometrically in a rectangular grid, so the variables in a time series are arranged in a one-dimensional linear array.

There is a large body of literature concerning the modeling of time series (see for example Kirchgässner & Wolters 2007 or Kantz & Schreiber 2004). A commonly used method for stationary times series, where the means and variances of the variables do not change over time, is the *autoregressive moving average* (ARMA) model, which we now describe. The assumption is that we generate a sequence of observations $\{X_i\}_i$, say for $i \geq 0$, and that X_i is the observation obtained at a time t_i. It is also assumed that the t_i are regularly spaced in time. The model further assumes that the observations at times t_i depend linearly on a fixed number (say p) of the preceding observations and also that there are models for noise in the system. Formally, one writes

$$X_t = c + \varepsilon_t + \sum_{i=1}^{p} \varphi_i X_{t-i} + \sum_{i=1}^{q} \theta_i \varepsilon_{t-i}.$$

Here the variables ε_i are typically assumed to be independent identically distributed (i.i.d.) random variables sampled from a normal distribution with zero mean. Models other than normal ones are possible, but it is required that the variables should still be i.i.d. Because of the presence of the random variables ε_t, the above model is not amenable a standard linear regression model but, rather, describes a distribution for each variable X_t. The fitting of the model is then carried out using a method for fitting distributions, a common choice being the *maximum likelihood* method (see Hastie et al. 2009). There are simple extensions to vector-valued time series. Another extension is the so-called *autoregressive integrative moving average* (ARIMA) model, which does not require that the time series be stationary.

There are of course situations where such linear models are not appropriate, but one nevertheless wants to predict the value of an observation using the preceding k observations for a fixed positive integer k. There is a construction that is available for time series, called *delay embedding* or *Takens embedding*.

To introduce this, we first consider how to construct dissimilarity spaces out of sequences from a dissimilarity space. Let $(X, \mathcal{D})$ be a dissimilarity space. We can form a dissimilarity space structure on the space of k-tuples $X(k) = (X^k, \mathcal{D}_k)$ where $\mathcal{D}_k$ can be chosen in a number of different ways. The most popular ones are in direct analogy with L_p-metrics on sequences:

$$\mathcal{D}_k((x_1, \ldots, x_k), (y_1, \ldots, y_k)) = \left(\sum_{i=1}^{k} \mathcal{D}(x_i, y_i)^p \right)^{1/p}.$$

Choosing $p = 1$ gives a particularly simple-to-use dissimilarity – just sum up the pairwise dissimilarities of the two sequences – while $p = 2$ is a very popular choice since in combination with the Euclidean metric it reduces to the Euclidean metric on concatenation of the observations.

Now, for a single time series $t = \{x_i\}_{i=1}^{N}$ we can define a *delay embedding* with delay ϵ and dimension k as the new time series

$$T_{k,\epsilon} t = \{(x_{i-(k-1)\epsilon}, \ldots, x_{i-\epsilon}, x_i)\}_{i=1+(k-1)\epsilon}^{N} \subseteq X(k).$$

If $\mathfrak{T} \subseteq X^N$ is a collection of time series samples, where each $t = (x_1, \ldots, x_N) \in \mathfrak{T}$ is a single time series taken over a family of times $(t_1, \ldots, t_N)$ we may generate a new collection of time series by applying delay embedding separately on each time series in the collection, producing a delay-embedded collection $T_{k,\epsilon} \mathfrak{T} = \{T_{k,\epsilon} t \mid t \in \mathfrak{T}\}$.

We will meet the Takens embedding again in Section 6.4, where the point cloud obtained by forgetting the time-ordering structure on $T_{k,\epsilon} \mathfrak{T}$ can be analyzed using persistent cohomology to generate information about quasiperiodicity in the time series itself.

2.8   Density Estimation in Point Cloud Data

The theory of density estimation is a highly developed area of statistics. We will not attempt to survey this area but will instead refer the reader to Scott (2015) or Duvroye

(1987). We will discuss a proxy for density that we have found useful and that will occur in some of the case studies.

One way to understand dense points is that the distance from the point to its kth neighbour is taken to be very small. Thus, in Figure 2.2, on the left, we can see that the distance to its 30th nearest neighbour is significantly smaller for a point in the denser region of the data set. We write $\nu_k(x)$ for the kth nearest neighbour of a point x, assuming the data set with which we are working to be known. Then we can define the *codensity* $\delta_k(x)$ by $\delta_k(x) = d(x, \nu_k(x))$. This quantity will rise as the density goes down; we can acquire a *density* estimator by working instead with $1/\delta_k(x)$ or we can work with the codensity directly and pick points with low values instead of points with high values for the dense subsets.

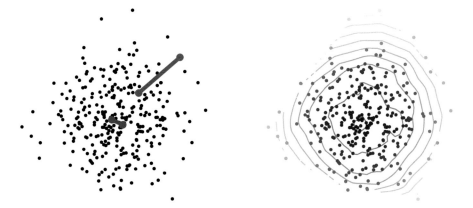

Figure 2.2 Three hundred points with a Gaussian bivariate distribution with variance 1.0 in each axis direction. Left: In red, top right, a point with high codensity, connected to its 30th nearest neighbour; in blue, a point with low codensity also connected to its 30th nearest neighbour. Right: The codensity δ_{30} of the points by color code.

It is worth noticing that different parameters k for the codensity give different structures on the resulting estimates. The right-hand part of Figure 2.2 is for $k = 30$. In Figure 2.3, the two figures show the corresponding functions for $k = 5$ and for $k = 200$.

The codensity climbs much more sharply in the low-parameter case, with almost all points sitting at a very low value and the ascent to high values of codensity being sudden as we arrive at the outliers. In comparison, for the high-parameter case, the ascent is more gradual, and medium to high codensity values appear at points much closer to the core.

In general, the higher the number of neighbours taken into account, the smoother the behaviour, and the more global the features that can be detected.

This gives us a way to deal numerically with *density*. It remains to consider just points of high density. For this, we introduce a second parameter. For a set $\mathcal{P}$, we write $\mathcal{P}(k, T)$ for the $T\%$ densest points in $\mathcal{P}$. That is, assuming no two points have exactly the same codensity, we order all points in $\mathcal{P}$ by codensity and pick the $\frac{T}{100}|\mathcal{P}|$ points with the lowest codensity values.

Figure 2.3 Points as in Figure 2.2. Left, codensity δ_5. Right, codensity δ_{200}. The 25% densest values are marked in red, indicating the sets $\mathcal{P}(5, 25)$ and $\mathcal{P}(200, 25)$ respectively.

For a subset $\mathcal{P}_0 \subset \mathcal{P}$, with a slight abuse of notation we shall write $\mathcal{P}_0(k, T)$ for the $\frac{T}{100}|\mathcal{P}_0|$ points of $\mathcal{P}_0$ on which the codensity – as calculated on all of $\mathcal{P}$ – takes on the lowest values.

Part II

Theory

3 Topology

3.1 History

Euler (1741) is usually cited as the first paper in topology. In this paper, Euler studies the so-called "Bridges of Königsberg" problem.

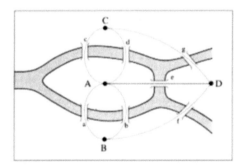

The question that was asked about the bridges was whether it is possible to traverse all the bridges exactly once and return to one's starting point. Euler answered this by recognizing that it is a question about paths in an associated network; see below. In fact, the question involves only certain properties of the paths which are independent of the rates at which the paths are traversed. His result concerned the properties of an infinite class of paths, i.e. a certain type of pattern in the network. Euler also derived a *polyhedral formula* relating the number of vertices, edges, and faces in polyhedra (Euler 1752a, 1752b). The subject developed in a sporadic fashion over the following century and a half; developments included the work of Vandermonde on knot theory (Vandermonde 1771), the proof of the Gauss–Bonnet theorem (never published by Gauss, but with a special case proved in Bonnet 1848), the first book in the subject, by Listing (1848), and the work of Riemann identifying the notion of a manifold (Riemann 1851). The paper of Poincaré (1895) was a seminal work in which the notions of homology and the fundamental group were introduced, with motivation from celestial mechanics. The subject then developed at a greatly accelerated pace throughout the twentieth century. The first paper in persistent homology was Robins (1999), and the subject of applying topological methodologies to finite metric spaces has been developing rapidly since that time.

3.2   Qualitative and Quantitative Properties

3.2.1   Topological Properties

The notion of a topological property is an old one in mathematics. We first discuss what is meant by topological properties, and why they are important in data science.

In many problems in science, one can impose different coordinate systems on a single physical problem. For example, one can apply rigid motions (rotations and translations) to data in Euclidean space. This kind of coordinate change is, for example, ubiquitous in physics. It has the useful property that it preserves distances between points. Often, however, one needs to consider coordinate changes that are not rigid motions but are more complicated, in the sense that they distort the geometric properties (such as distance) of the data obtained from observation or experimentation.

Example 3.1   Temperature can be measured using many different scales, including the Celsius, Fahrenheit, and Kelvin scales. The transition from degrees Celsius to degrees Fahrenheit is given by the formula

$$F = \frac{9}{5}C + 32.$$

This transformation does not preserve distances; it dilates them by a constant factor. The transformation law from degrees Celsius to kelvins is given by

$$K = C + 273.15.$$

This transformation does preserve distances, i.e. the size of a degree.

Example 3.2   A change of coordinates which is frequently used to make apparent some properties of data is the so-called log–log coordinate system, in which every point (x, y) is replaced by $(\log(x), \log(y))$. This transformation carries curves given by power laws to straight lines of varying slopes. It does not preserve distances.

Example 3.3   Polar coordinates provide a convenient way to study a number of problems. The transition law from polar coordinates to rectangular coordinates does not preserve distances.

Example 3.4   Coordinate changes which apply multiplication by non-orthogonal matrices will distort the geometric properties of sets in the plane or in space. For example, dilation (multiplication by a positive multiple of the identity matrix) carries a circle centered at the origin to another circle centered at the origin with a different radius. Multiplication by a general diagonal matrix (perhaps with distinct eigenvalues) carries circles centered at the origin to ellipses centered at the origin.

Example 3.5   It is often said that topology is the subject in which a coffee cup is regarded as being the same as a doughnut. This means that there is a coordinate change such that a set which is a coffee cup in one coordinate system is a doughnut in the new coordinates. See the illustration below.

Example 3.6 Complicated coordinate changes can transform the representation of a letter given in one font to the same letter in the new font.

Similarly, a coordinate change can carry the letter "C" to the letter "I".

Properties which are not altered by arbitrary continuous changes of coordinates are called *topological*. Below some examples of topological properties are given. At the moment, we are thinking intuitively, not formally. Formal definitions of these properties will come in due course.

Example 3.7 For X a subset of the plane, we may ask what is the number of connected pieces into which it breaks. A connected piece is a subset $Y \subseteq X$ such that we can connect every pair of points in Y with a path *contained in Y*. This number is clearly independent of the coordinate system used to study X. This idea extends trivially to subsets of higher-dimensional $\mathbb{R}^n$.

Example 3.8 For X a subset of the plane, we may ask how many holes it has. A useful way to think about this is to recognize that the presence of holes corresponds to the presence of loops in X, and that in an appropriate sense we may count the number of *independent loops* in X, suitably defined. This number of independent loops is also a topological property. For the figure below, this number is 3.

Example 3.9 For X again a subset of the plane, we may ask how many *ends* X has. The figure below has four ends.

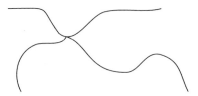

There are many other such properties, and we will be performing a systematic construction of such properties. There are at least two different ways to apply topological properties in data science.

The first is as a tool for gaining an understanding of the large-scale structure of data sets and for the related task of producing simplifying models for them. Informally, we say that a data set $\mathbb{D} \subseteq \mathbb{R}^n$ is *modeled by a subset* $X \subseteq \mathbb{R}^n$ if in an appropriate sense all points of $\mathbb{D}$ are close to X. It might be that the points appear to be sampled, with noise, from X. For example, when performing linear regression we might find that the data is concentrated very near to a line. The topological properties of a model space X can give useful cues that suggest that particular kinds of phenomena are occurring within the data and can also suggest good synthetic models for $\mathbb{D}$, as is done in linear regression.

1. If we find that X breaks up into several connected pieces, then this suggests that there is an appropriate taxonomy of $\mathbb{D}$ into a number of taxonomic units, one for each connected piece of X. Such a taxonomy is a kind of model for the data, and often yields great understanding of it. It can also allow for the construction of classifiers that can identify to which taxonomic unit a new point belongs.
2. If we find that X has a loop in its structure, this suggests that the data is derived from periodic or recurrent behavior. The data can then often be viewed as moving around the loop with time.
3. If we find that X has ends, they can often be interpreted as different modes of anomalous behavior, with the points away from the nodes interpreted as the normal modes of behavior. For example, suppose that our data has been collected from various sensors aboard an airliner; then we might consider four distinct modes, namely (1) flying at altitude in non-turbulent conditions, (2) taking off, (3) landing, and (4) flying at altitude in turbulent conditions. This kind of data would be expected to have a model space which looks like a letter "Y", with the central node where all the spokes intersect corresponding to situation (1), and the tips of the spokes corresponding to situations (2), (3), and (4).
4. If we find that X has bubbles in its structure, including "0-dimensional" (disconnected) pieces or "one-dimensional" loops, these can suggest that the data has a *missing middle*: that for some reason there is a combination of features that is absent in the data. We will see an example of this in Section 6.9, where in voting patterns in the US House of Representatives we can observe an absence of bills that have only partial support from each of the parties. Most bills that come to a vote have at least one party fully backing them and this shows up as a topological feature in the data.

Remark 3.10   Although we have used the notion of a model space for a data set $\mathbb{D}$, it turns out that we do not have to assume the existence of such a model in order to be able to identify topological properties of interest. Much of this volume is devoted to the study of methods that can attach topological properties to data sets directly, without having to construct an explicit space X.

The second application of topological properties to data science is as a method for *feature generation*. Data scientists often distinguish between "structured" and "unstructured" data. Structured data typically means data that comes in the form of a matrix, a table in a database, or a spreadsheet, all with numerical entries; unstructured data means all other kinds of data. Often, though, apparently unstructured data are actually equipped with some kind of structure, which is different from the standard structures described above. For example, databases of molecules might be equipped with a structure in which each data point comprises a list of atoms and positions, together with the bonds connecting the atoms. This kind of data is difficult to represent effectively in a matrix or spreadsheet format. There are numerous reasons why this is so. For example, the number of atoms may vary from molecule to molecule, as may the number of bonds. Also, in order to represent molecules using any kind of matrix, we must choose an ordering on the set of atoms and bonds, which is entirely artificial since there is no assumed ordering on them "in nature". This is illustrated by the molecular bold diagram below, in which there is no obvious natural ordering on the atoms.

The fact that unstructured data is not in matrix form is a problem for data scientists, since most algorithms in machine learning rely on matrix representations. What this means is that methods for the *feature generation* of unstructured data are of great importance. Of course, the creation of such features depends in a fundamental way on any structure present in the data. For the case of databases of molecules, each data point (i.e. molecule) itself has geometric or topological properties, such as loops and ends. If we are able to identify such features and quantify them, we can treat each such value as a feature to be attached to the molecule, and these features can then be used as entries in a data matrix attached to the molecules. We will see applications of this idea in Section 6.2. Databases of images are also amenable to this kind of topological feature generation.

Ultimately we want to make precise mathematical and computational sense of the topological properties about which we have talked. In order to do this, we now define the notion of a *topological space*, which defines the kind of objects on which topological properties can be studied. A starting point is $\mathbb{R}^n$, but we will find that it is very useful to regard a topological space as a generalization of $\mathbb{R}^n$ on which we can define familiar notions, such as continuous maps. The idea is as follows. In $\mathbb{R}^n$ we often study the properties of the distance function. It can be used to define continuous functions, since the usual $\epsilon - \delta$ definition of continuity uses only the distance function. However, we find that arbitrary continuous transformations can distort distances in a very flexible way; it is not possible to define topological properties based on transformations that preserve the distance function. One wants, in a sense, to preserve "infinitesimal" distances, i.e. points which are infinitely close together should remain infinitely close together. Of course, there is no notion of infinitesimal closeness for pairs of points. However, in a metric space one can speak of a point being infinitesimally close to a set. For (X, d) a metric space, $x \in X$, and $U \subseteq X$, we say that x is *infinitesimally close* to U if, for every $\epsilon > 0$, there is a point in $u \in U$ such that $d(x, u) \leq \epsilon$. The collection of all points which are infinitesimally close to a set U is called the *closure* of U and is denoted by $\overline{U}$. Note that if a point is infinitesimally close to a set U then any continuous transformation will preserve that property, even if it does not preserve distances.

Example 3.11   The closure of the open interval $(0, 1)$ is the closed interval $[0, 1]$.

Example 3.12   The closure of the set $\mathbb{R}^n - \{0\}$ in $\mathbb{R}^n$ is the entire set $\mathbb{R}^n$.

Example 3.13   The closure of the set of all points of the form $\frac{n}{2^k}$, where $0 \leq n \leq 2^k$ and $k \geq 0$, is the entire closed interval $[0, 1]$.

Closure can be viewed as an operator on the subsets of $\mathbb{R}^n$. It turns out to be *idempotent* in the sense that $\overline{\overline{U}} = \overline{U}$. One can develop the notion of a topological space using only an abstract closure operator on the family of subsets, with certain properties. However, it is more conventional to develop the notion using the ideas of closed and open sets. A *closed* set V in $\mathbb{R}^n$ is a subset for which $\overline{V} = V$. A subset U of $\mathbb{R}^n$ is *open* if its complement in X is closed, i.e. if $U = X - V$ for some closed set V. Another characterization is that a set is open if and only if it is an arbitrary union of open balls $B_r(x) = \{y \in \mathbb{R}^n \mid d(x, y) < r\}$. The open sets in a metric space $\mathbb{R}^n$ satisfy the following three properties.

- $\emptyset$ and X are both open sets in X.
- Arbitrary unions of open sets are open.
- Finite intersections of open sets are open.

The identification of these properties leads us to the notion of a topological space.

Definition 3.14   A *topological space* is a pair $(X, \mathcal{U})$, where X is a set and $\mathcal{U}$ is a family of subsets (called the open sets) of X satisfying the following three properties.

1. $\emptyset \in \mathcal{U}$ and $X \in \mathcal{U}$.
2. For any family $\{U_\alpha\}_{\alpha \in A}$ of open sets, where A is any parameter set, $\bigcup_{\alpha \in A} U_\alpha$ is also in $\mathcal{U}$.
3. For any family $\{U_\alpha\}_{\alpha \in A}$ of open sets, where A is any finite parameter set, $\bigcap_{\alpha \in A} U_\alpha$ is also in $\mathcal{U}$.

The family $\mathcal{U}$ is called a *topology* on X. A subset $C \subseteq X$ is *closed* if $X \setminus C$ is open.

Here are some examples of open and closed sets in $\mathbb{R}^n$.

Example 3.15   Open (respectively closed) intervals are open (respectively closed) sets in $\mathbb{R}$.

Example 3.16 Open balls (respectively closed balls) are open (respectively closed) sets in $\mathbb{R}^n$.

Example 3.17 Finite sets of points in $\mathbb{R}^n$ are always closed. Complements of finite sets are open.

Example 3.18 For continuous function $f: \mathbb{R}^n \to \mathbb{R}$, the set of points $\{v \in \mathbb{R}^n \mid f(v) < 0\}$ is always an open set. Similarly for $>$. The set of points $\{v \in \mathbb{R}^n \mid f(v) \leq 0\}$ is always a closed set. Again, similarly for $\geq$. The terminology is consistent with the notions of open and closed conditions used in analysis.

Example 3.19 For any family $\{f_1, f_2, \ldots, f_k\}$ of continuous functions from $\mathbb{R}^n$ to $\mathbb{R}$, the set $\{v \in \mathbb{R}^n \mid f_i(v) = 0$ for all $i\}$ is closed. This means that algebraic varieties (sets of points defined by algebraic equations) are closed sets in $\mathbb{R}^n$.

As we progress, we will give numerous examples of topological spaces. Here we introduce the most elementary example.

Definition 3.20 For any subset $X \subseteq \mathbb{R}^n$, we define a topology on X by declaring that $U \subseteq X$ is open if and only if $U = V \cap X$ for an open subset $V \subseteq \mathbb{R}^n$. This defines familiar spaces such as spheres and tori as topological spaces.

We are now able to be precise about what is meant by a topological property.

Definition 3.21 A property of subsets of $\mathbb{R}^n$ is said to be *topological* if and only if it depends only on the topology defined above. Of course, topological properties can be defined on other topological spaces as well.

The definition of a topological space is motivated by the case of $\mathbb{R}^n$, and so clearly applies to it. The power of the definition lies in the fact that there are a number of systems far removed from the case of $\mathbb{R}^n$ but which nevertheless satisfy the axioms for a topological space. It turns out that this enables the use of geometric and topological intuition in cases where it was not anticipated.

Example 3.22 For any set X, we can let the family of all subsets of X with finite complement be the open sets. This does give a topological space.

Example 3.23 Let p be a prime number. A topology $\mathcal{U}$ on $\mathbb{Z}$, the set of integers, is given by defining $\mathcal{U}$ to be the collection of sets U such that for any $n \in U$ there is a positive integer r such that the set $\{n + kp^r \mid k \in \mathbb{Z}\}$ is contained in U. This is called the p-adic topology, and it is of use in number theory.

3.2.2 Continuous Maps and Homeomorphisms

In the study of multivariable calculus, we identify the notion of continuous maps from $\mathbb{R}^m$ to $\mathbb{R}^n$ using the ϵ–δ definition of continuity. We can characterize continuity directly using the topology on $\mathbb{R}^n$ as follows.

Proposition 3.24   *A map $f: \mathbb{R}^m \to \mathbb{R}^n$ is continuous if and only if $f^{-1}(U)$ is open for all open sets $U \subseteq \mathbb{R}^n$.*

This leads us to the notion of a continuous map between topological spaces X and Y.

Definition 3.25   Let X and Y be topological spaces and $f: X \to Y$ a map of sets. Then f is said to be continuous if and only if $f^{-1}V$ is open for all open sets $V \subseteq Y$.

Remark 3.26   One can formulate the notion of a closure operator in an arbitrary topological space and then interpret this definition as requiring that if $x \in X$ is in the closure of a subset $U \subseteq X$ then $f(x)$ is in the closure of $f(U)$.

Using this notion of maps, we will now identify the key notion that will allow us to speak precisely about topological properties.

Definition 3.27   Let $f: X \to Y$ be a continuous map of topological spaces. If there is a continuous function $g: Y \to X$ such that these functions are inverses of each other, i.e. $f(g(y)) = y$ and $g(f(x)) = x$, then f is a *homeomorphism*, and we say that X and Y are *homeomorphic*.

Remark 3.28   This notion is the exact analogue of the notion of the isomorphisms of groups or the bijections of sets. It means that the map f can be inverted and that the inverse is continuous. It also gives precise meaning to the idea that Y can simply be regarded as a reparametrization of X. Another point of view is that it makes precise the notion of Y being obtained from X by stretching or deforming, without tearing or "crushing". If there is a homeomorphism from X to Y, we say that X and Y are homeomorphic.

Here are examples of pairs of homeomorphic spaces.

Example 3.29   Let $X \subseteq \mathbb{R}^2$ denote the unit circle, and let Y denote the ellipse given by the equation

$$\frac{x^2}{a^2} + \frac{y^2}{b^2} = 1.$$

Both sets are equipped with the subspace topology, defined in Definition 3.20. Then the map $(x, y) \to (ax, by)$ is a homeomorphism from the circle to the ellipse, so the circle and the ellipse are homeomorphic.

Example 3.30   The parabola given by $y = x^2$ and the real line are homeomorphic, the homeomorphism being given by the map $(x, y) \to x$.

Example 3.31   Let $X \subseteq \mathbb{R}^2$ denote the set $X = \{(x, y) \in \mathbb{R}^2 \mid x^2 + y^2 > 1\}$, and let $Y \subseteq \mathbb{R}^2$ denote the subset $\mathbb{R}^2 - \vec{0}$. Polar coordinates are a good choice for describing a map $g: X \to Y$, which we define by

$$g(r, \theta) = (r - 1, \theta).$$

The inverse map is given by

$$g^{-1}(r, \theta) = (r + 1, \theta).$$

Example 3.32 Let X be the torus in three-dimensional space defined by the equation

$$(c - \sqrt{x^2 + y^2})^2 + z^2 = a^2,$$

where $0 < a < c$. Let $Y \subseteq \mathbb{R}^4$ be the subspace defined by the equations

$$\begin{cases} x^2 + y^2 = 1, \\ z^2 + w^2 = 1. \end{cases}$$

We define a map $f : X \to Y$ via the formula

$$f(x, y, z) = \left(\frac{x}{\|(x, y)\|_2}, \frac{y}{\|(x, y)\|_2}, \frac{c - \|(x, y)\|_2}{a}, \frac{z}{a} \right),$$

where $\|(x, y)\|_2$ denotes $\sqrt{x^2 + y^2}$. It is easy to verify that this map has its image in Y, and it is continuous since it is defined by simple formulas. It turns out that f is a homeomorphism.

One way to prove that the map f in the above example is a homeomorphism is to exhibit a formula for the inverse. This is somewhat involved even in this case and can become much more so in more complicated examples. There is a simple statement that allows one to bypass that process.

Proposition 3.33 *Let $f : X \to Y$ be a continuous map, where X and Y are subsets of $\mathbb{R}^m$ and $\mathbb{R}^n$ respectively. We say a subset of $\mathbb{R}^n$ is* bounded *if it is contained in a ball $B_R(0) = \{v \mid \|v\| < R\}$ for some $R > 0$. Suppose that X and Y are closed and bounded. If the map f is bijective as a map of sets, then f is a homeomorphism.*

In Example 3.32 above, it is easy to check that both sets are closed and bounded and that f, regarded as a map of sets, is bijective. We are therefore able to conclude that it is a homeomorphism.

Remark 3.34 Note that the two descriptions of a topological space have different strengths. The first description sits in three-dimensional space and can therefore be readily visualized. The second is much simpler, in that the equations that define it are much simpler. Depending on the kind of analysis one is doing, different coordinate systems may be useful.

We should also ask how we can show that two spaces are *not* homeomorphic. Let $X = [0, 1] \subseteq \mathbb{R}$ and let $Y \subseteq \mathbb{R}^2$ denote the closed unit ball $\{(x, y) \in \mathbb{R}^2 \mid x^2 + y^2 \leq 1\}$. These two spaces appear to be very different and we therefore suspect that they are not homeomorphic, but in order to be convinced of this we must try to find some topological property that applies to one but not to the other. Reasoning intuitively, we observe that by removing a single point from X, such as the point $\frac{1}{2}$, we can form a space which is disconnected in the sense that is breaks into two disjoint open pieces, namely $[0, \frac{1}{2})$ and $(\frac{1}{2}, 1]$, while removing any single point from Y will always leave it in a single connected piece. We have not formally defined connectedness yet (we will do so in Section 3.2.8), but this informal argument suggests how we can develop topological properties that enable discrimination between spaces, even when we allow ourselves to modify the spaces by arbitrary coordinate changes.

3.2.3   Metric Spaces

Euclidean space $\mathbb{R}^n$ is equipped with a distance function, given by

$$d(\vec{x}, \vec{y}) = \sqrt{(\vec{x} - \vec{y}) \cdot (\vec{x} - \vec{y})}.$$

Consequently, we have a precise way to measure how close points are to each other. This means that, for any subset $X \subseteq \mathbb{R}^n$, we have a distance function obtained by restricting the distance function on Euclidean space to X. In mathematics, one of the most important instances of this kind of space is solution sets to equations and inequalities, or to systems of equations and inequalities. Here are some of the most frequently occurring examples.

Example 3.35 $D^n \subseteq \mathbb{R}^n$ is defined to be the set

$$\{\vec{x} \mid \vec{x} \cdot \vec{x} \leq 1\}.$$

It is called the *unit disc* in $\mathbb{R}^n$.

Example 3.36   The *unit sphere* in $\mathbb{R}^{n+1}$, denoted by $S^n \subseteq \mathbb{R}^{n+1}$, is the set

$$\{\vec{x} \mid \vec{x} \cdot \vec{x} = 1\}.$$

Example 3.37   The *standard n-simplex*, written $\Delta^n \subseteq \mathbb{R}^{n+1}$, is the set

$$\left\{ (x_0, x_1, \ldots, x_n) \mid \sum_i x_i = 1 \ \text{ and } x_i \geq 0 \text{ for all } i \right\}.$$

Example 3.38   The *embedded torus with radii R and* ρ, with $\rho < R$, is the surface of revolution obtained by rotating a circle of radius ρ in the xz-plane centered at the point $(R, 0)$ around the z-axis. It can be parametrized using coordinates (θ, φ) via the formulae

$$\begin{cases} x = (R + \rho \cos \varphi) \cos \theta, \\ y = (R + \rho \cos \varphi) \sin \theta, \\ z = \rho \sin \varphi. \end{cases}$$

We have already discussed this space in Example 3.32.

It is possible to perform a number of constructions on subsets of Euclidean space to obtain new spaces.

Definition 3.39   Let $X, Y \subseteq \mathbb{R}^n$. Then, by $X \coprod Y$, the *disjoint union* of X and Y, we mean the subset

$$\{(\vec{x}, 0) \mid \vec{x} \in X\} \cup \{(\vec{y}, 1) \mid \vec{y} \in Y\} \subseteq \mathbb{R}^{n+1}.$$

Definition 3.40   Let $X \subseteq \mathbb{R}^m$ and $Y \subseteq \mathbb{R}^n$. Then by $X \times Y$, the *product* of X and Y, we mean the subset

$$\{(\vec{x}, \vec{y}) \in \mathbb{R}^{m+n} \mid \vec{x} \in X \text{ and } \vec{y} \in Y\}.$$

Definition 3.41 Let $X \subseteq \mathbb{R}^n$ and $r \in \mathbb{R}$. Then by the *cone* on X with apex r, we mean the subset $C_r X \subseteq \mathbb{R}^{n+1}$ defined by

$$C_r X = \left\{ (\vec{x}, t) \mid 0 \le t \le 1 \text{ and } (\vec{x}, t) = (t\vec{\xi}, (1-t)r) \text{ for some } t \text{ and some } \xi \in X \right\}.$$

The *suspension* of X, denoted by ΣX, is the union of the two cones $C_1(X)$ and $C_{-1}(X)$.

Example 3.42 For any closed set $X \subseteq \mathbb{R}^n$, we can define the *one-point compactification* of X to be $X \cup \{\infty\}$, a subset of $X \cup \{\infty\}$ being open if either (a) it is an open subset of X or (b) it is of the form $U \cup \{\infty\}$, where U is an open subset of X containing a set of the form $\{v \in \mathbb{R}^n \mid |v| > R\}$ for some $R > 0$.

We have used the notion of a distance function to describe a notion of closeness for pairs of points in Euclidean space, and therefore for points in subsets of Euclidean space. However, it is interesting and useful to abstract the most important properties of this distance function and develop a more general notion of distance function. This will give new spaces which either cannot be embedded as subsets of Euclidean space or where the choice of such an embedding is artificial. We will call them *metric spaces*. A metric space is a pair (X, d), where X is a set and where $d \colon X \times X \to [0, +\infty)$ is a function. The key properties of the Euclidean distance function which we also require of d are the following.

- **Nonnegativity:** $d(x, y) \ge 0$ for all x, y, and $d(x, y) = 0$ if and only if $x = y$.
- **Symmetry:** $d(x, y) = d(y, x)$.
- **Triangle inequality:** $d(x, z) \le d(x, y) + d(y, z)$ for all x, y, z.

For any $r > 0$ and $x \in X$, we will denote by $B_r(x)$ the set $\{x' \in X \mid d(x, x') < r\}$. We will refer to it as the *open ball of radius r centered at x*.

Of course, the Euclidean distance function and its restriction to any subset of Euclidean space satisfy these axioms. Here are some other examples.

Example 3.43 (Intrinsic distance) For any subset X of $\mathbb{R}^n$ that is path-connected, in the sense that, for any pair of points $x, x' \in X$, there is a continuous map $\varphi \colon [0, 1] \to X$ with $\varphi(0) = x$ and $\varphi(1) = x'$, we define the *intrinsic distance* from x to x' to be

$$\inf_{\varphi} \lambda(\varphi),$$

where φ ranges over all paths beginning at x and ending at x', and where λ denotes the arclength of φ. This idea leads to the notion of Riemannian manifolds, which have their own notion of distance.

Example 3.44 (Hamming distance) Let $\mathcal{B}^n$ denote *Boolean n-space*, the set of ordered n-tuples of elements of the set $\{0, 1\}$. Then the *Hamming distance* between vectors $\vec{x}$ and $\vec{y}$ is just the number of coordinates i such that $x_i \ne y_i$. This distance function can be modified by introducing weightings on each of the variables, so that

$$d^{\text{Hamming}}(\vec{x}, \vec{y}) = \sum_i \alpha_i \delta(x_i, y_i),$$

where α_i is a weighting vector; $\delta(x, y) = 0$ when $x = y$ and $\delta(x, y) = 1$ when $x \ne y$.

Example 3.45 (Graph distance)   Let $\Gamma = (V, E)$ denote a *graph* in the combinatorial sense: V is a set, and $E \subseteq V \times V$ is any subset with the symmetry property that $(x, x') \in E \Leftrightarrow (x', x) \in E$. An *edge path* in Γ is a sequence $(v_0, v_1, \ldots, v_n)$ such that $(v_i, v_{i+1}) \in E$ for all i. The number n is called the *length* of the edge path. We now define a distance function on V by declaring that the distance from v to w is the length of the shortest edge path beginning at v and ending at w.

Example 3.46 (Edit or Levenshtein distance)   Let A denote any alphabet, i.e. set, typically finite. We will construct a metric space whose underlying set is the collection of strings of elements in A. We say that two such strings α, β are *adjacent* if β can be obtained from α via one of the following "moves".

- Substitute any element of A for one of the symbols in α.
- Delete a symbol of α.
- Insert any element of A between any two consecutive symbols in the string α.

We then define the *edit* or *Levenshtein* distance between two strings σ and τ to be the minimal integer n such that there is a sequence of strings $(\sigma_0, \sigma_1, \ldots, \sigma_n)$ for which

- $\sigma_0 = \sigma$,
- $\sigma_n = \tau$,
- σ_i and σ_{i+1} are adjacent for all $i = 0, \ldots, n - 1$.

This metric plays a key role in the bioinformatic analysis of DNA and protein sequences. Although it might appear difficult to compute, algorithms have been developed which make it computationally tractable.

Example 3.47   We say a function $f : \mathbb{R} \to \mathbb{R}$ is *convex* on an interval $[a, b]$ if, for any $x, y \in [a, b]$ and any $0 < \lambda < 1$, we have that

$$f(\lambda x + (1 - \lambda)y) \leq \lambda f(x) + (1 - \lambda)f(y).$$

Given a metric space (X, d) and a convex function f on $[0, +\infty]$, one can observe that the function $f \circ d : X \times X \to \mathbb{R}$ also defines a metric on X.

Example 3.48 (Cosine similarity metric)   Consider the unit sphere

$$S^n = \left\{ \vec{x} \in \mathbb{R}^{n+1} \mid \vec{x} \cdot \vec{x} = 1 \right\}.$$

Then the function $d(\vec{x}, \vec{y}) = 1 - \cos^{-1}(\vec{x} \cdot \vec{y})/\pi$ defines a metric on S^n. Note that the dot product is just the cosine of the angle between the vectors.

Metric spaces are important for us because a metric on a set X allows us to construct a topology on X.

Definition 3.49   Let (X, d) denote a metric space. Then the *topology associated with d* is the family of subsets that are arbitrary unions of open balls $B_r(x)$. It is easy to check that this family is a topology.

Remark 3.50 This tells us that selecting a metric on a set X (which is generally much simpler than specifying a topology directly) gives a method for constructing topological spaces.

Consider the following physical situation. Suppose that a planet is orbiting around a star, and that it is being observed from two separate locations, one on Earth, and one on a space telescope. We will suppose that both observation posts are recording the position of the planet. Each observation post will record the data in a coordinate system which is centered at its telescope, and in which the z-axis is the straight line from the telescope to the star. We assume that the coordinate systems are such that the axes are all perpendicular to each other, and also that all distances are measured in kilometers in both systems. Each observation post now produces a set of points in Euclidean 3-space, by recording the positions of the planet in its coordinate system. We thus obtain two different sets of points in $\mathbb{R}^3$, representing the same set of positions of the planet. However, note that by performing a rigid coordinate change (a rigid motion, i.e. a composition of a translation and a three-dimensional rotation, which preserves distances and angles), we can obtain one set from the other. Scientists are interested in the inherent physical properties of the system, not artifacts arising from the details of the measurement such as the choice of coordinates. For this reason, they are very interested in the properties of the system which do not change once a rigid coordinate change is applied. This kind of invariance is sufficiently important for us to generalize it to metric spaces.

Definition 3.51 Let (X, d_X) and (Y, d_Y) be metric spaces. A set map $f : X \to Y$ is said to be an *isometry* it is bijective, and if, for every pair of points $x, x' \in X$, we have that

$$d_X(x, x') = d_Y(f(x), f(x')).$$

If there is an isometry from X to Y, we say that X and Y are *isometric*.

Isometries thus preserve distances exactly. We can regard two isometric metric spaces as effectively indistinguishable from each other, since any metric information in one space can be recovered from the other, and therefore one can be regarded as obtained from the other by simply renaming the points. For subsets of Euclidean space, the statement that two subsets are isometric can frequently be interpreted as saying that one can be obtained from the other by applying rigid motions. In many applications, sets in which we are interested come equipped with choices of coordinate systems in which many arbitrary choices may have been made. Most of the properties and quantities in which we are interested will be independent of the choice of coordinates. This means that these properties and quantities will not be altered by the application of an isometry. We have the following.

Proposition 3.52 *If $f : (X, d_X) \to (Y, d_Y)$ is an isometry of metric spaces then f is a homeomorphism between the associated topological spaces.*

We will also need to study the notion of continuous maps between metric spaces. Given two metric spaces X and Y, a set map $f : X \to Y$ is informally thought of as being *continuous* if it "carries nearby points to nearby points". The mathematically precise version of this idea is the following.

Definition 3.53   Let (X, d_X) and (Y, d_Y) be metric spaces. A set map $f: X \to Y$ is said to be *continuous at a point* $x \in X$ if for every $\epsilon > 0$ there is a $\delta > 0$ such that whenever $d_X(x, x') \leq \delta$ we have $d_Y(f(x), f(x')) \leq \epsilon$. The map f is said to be *continuous* if it is continuous at x for every $x \in X$.

This notion coincides with our usual intuitive notion for continuous maps between subsets of Euclidean spaces. Isometries are of course always continuous.

Proposition 3.54   *Let $f: (X, d_X) \to (Y, d_Y)$ be a map of metric spaces that is continuous at every point of X. Then f is a continuous map of the associated topological spaces.*

All the notions that we are developing, which are applied most commonly to subsets of Euclidean space, will still apply in the more general setting of metric spaces. We will find that this added generality is important.

3.2.4   Homotopy and Homotopy Equivalences

Suppose that we are studying a data set with two real-valued features, and which is concentrated around the unit circle in $\mathbb{R}^2$. We will suppose that the data comes with an error that is bounded in magnitude by a small number ϵ. What this means is that the data lie in the annulus A_ϵ described in polar coordinates by $A_\epsilon = \{(r, \theta) \mid 1 - \epsilon \leq r \leq 1 + \epsilon\}$. Since we do not have control over the error in the measurements, the shape we could expect to observe would be A_ϵ, not the unit circle, which would be observable only with infinite precision of the measurements. Next, suppose that we are able to make an additional measurement, so that the data now lies in $\mathbb{R}^3$. Suppose further that the new feature is not informative, e.g. that it is simply uniformly distributed with mean zero on a small interval. If we make these three measurements then what we would expect to observe is data concentrated on a solid torus $\mathbb{T}$ containing the unit circle in $\mathbb{R}^3$. Therefore, the space we can hope to observe from the three variables is this torus. From a data-analytic point of view, it is desirable that the results of our analysis should not depend on whether we measured in $\mathbb{R}^2$ or $\mathbb{R}^3$ since the third variable is not informative. However, the annulus and the torus are *not* homeomorphic, as one can readily check. For this reason, it is very useful to study a more flexible notion of equivalence than homeomorphism, which we call *homotopy equivalence*. The idea is to permit certain kinds of collapses of sets into points. For example, all the unit discs $D^n \subseteq \mathbb{R}^n$ are homotopy equivalent. We will first need to define the notion of a homotopy between two continuous maps.

Intuitively, a homotopy from f to g is a continuous one-parameter family of deformation maps, beginning at f and ending at g.

The figure shows a homotopy between two maps f and g from the circle S^1 into the Euclidean plane. Each of the trajectories corresponds to one member of the family of deformations, and is a map from S^1 to the plane. In order to formalize this into a mathematical definition, we will need to introduce the notion of a product of topological spaces.

Definition 3.55   Let X and Y be two topological spaces. By the product of X and Y, denoted by $X \times Y$, we will mean the space which is the set of ordered pairs (x, y), with $x \in X$ and $y \in Y$, and for which the open sets are arbitrary unions of sets of the form $U \times V$, where U and V are open sets in X and Y, respectively. It is readily verified that this family of sets forms a topology on $X \times Y$.

Example 3.56   The product $[0, 1] \times [0, 1]$ is homeomorphic to the set $\{(x, y) \in \mathbb{R}^2 \mid x \in [0, 1] \text{ and } y \in [0, 1]\}$.

Example 3.57   The product $S^1 \times [0, 1]$ is homeomorphic to the set $\{(x, y, z) \in \mathbb{R}^3 \mid x^2 + y^2 = 1 \text{ and } 0 \leq z \leq 1\}$; it corresponds to the vertical part of the boundary of a can. It is also homeomorphic to the annulus $\{(x, y) \in \mathbb{R}^2 \mid 1 \leq \sqrt{x^2 + y^2} \leq 2\}$.

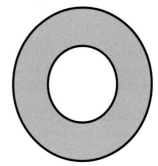

Example 3.58   The product $S^1 \times S^1$ of two copies of the circle is homeomorphic to the spaces defined (and shown to be homeomorphic) in Example 3.32. It is therefore homeomorphic to a torus.

We are now in a position to define precisely the notion of the homotopy of continuous maps.

Definition 3.59   Let $f, g \colon X \to Y$ be continuous maps from a space X to a space Y. By a *homotopy* from f to g we will mean a continuous map $H \colon X \times [0, 1] \to Y$ such that $H(x, 0) = f(x)$ and $H(x, 1) = g(x)$. As usual, $[0, 1]$ denotes the unit interval in $\mathbb{R}$. When there is a homotopy from f to g, we say that f and g are *homotopic* and we write $f \simeq g$.

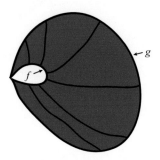

The figure shows the arcs obtained by restricting the map H to sets of the form $x \times [0, 1]$, where $x \in X$.

Here are some examples of homotopies.

Example 3.60   Suppose that a space X consists of a single point. Then a continuous map from X to Y consists simply of choosing a point y in Y. The maps associated with the points y and y' are homotopic precisely if there is a continuous map φ from $[0, 1]$ into Y such that $\varphi(0) = y$ and $\varphi(1) = y'$. The intuitive interpretation is that there is a *path* in Y connecting the two points.

Example 3.61   Suppose that a target space Y is $\mathbb{R}^n$, and let $f, g \colon X \to Y$ be any two continuous maps. Then the map $H \colon X \times [0, 1] \to Y$, given by

$$H(x, t) = (1 - t)f(x) + tg(x),$$

is a homotopy from f to g, where the addition operator is taken to be the usual vector addition in $\mathbb{R}^n$. This *straight-line homotopy* works whenever the target space is convex. In other words, any two maps into Euclidean space are homotopic.

Example 3.62   Consider two maps from the punctured complex plane $\mathbb{C}^* = \mathbb{C} - \{0\}$ to itself, given by $f(z) = z^n$ and $g(z) = 2z^n$. Then there is a homotopy from f to g given by

$$H(z, t) = e^{\ln 2t} z^n.$$

Note that because we are constructing a homotopy of maps to the space $\mathbb{C}^*$, we must make sure that the map H never takes the value 0.

Remark 3.63   If we take the two maps $f(z) = z^m$ and $g(z) = z^n$, where $m \neq n$, it turns out that f and g are *not* homotopic as maps from $\mathbb{C}^*$ to itself, although we are not yet in a position to demonstrate this.

Definition 3.64   Let $f: X \to Y$ be a continuous map. We say that f is a *homotopy equivalence* if there is a continuous map $g: Y \to X$ such that $f \circ g$ is homotopic to the identity map on Y and $g \circ f$ is homotopic to the identity map on X. Two spaces are *homotopy equivalent* if there is a homotopy equivalence from one to the other. When there are fixed base points, $x_0 \in X$ and $y_0 \in Y$, we speak of *based maps* as maps $f: X \to Y$ for which $f(x_0) = y_0$, and of based homotopies as homotopies $H(x_0, t) = y_0$ for all t, and we thus have the analogous notion of *based homotopy equivalence*.

Evidently, if two spaces are homeomorphic, they are homotopy equivalent. On the other hand, there are many situations where spaces are homotopy equivalent but not homeomorphic.

Example 3.65   The inclusion $\{\vec{0}\} \hookrightarrow \mathbb{R}^n$ is a homotopy equivalence. In other words, the one-point space and $\mathbb{R}^n$ are homotopy equivalent. Explicitly, the composite

$$\{\vec{0}\} \hookrightarrow \mathbb{R}^n \longrightarrow \{\vec{0}\}$$

is equal to the identity, and the composite

$$\mathbb{R}^n \longrightarrow \{\vec{0}\} \hookrightarrow \mathbb{R}^n$$

is homotopic to the identity via the homotopy

$$H(\vec{v}, t) = tv.$$

It is clear that the two spaces are not homeomorphic, since $\mathbb{R}^n$ consists of more than one point.

A space that is homotopy equivalent to the one-point space is called *contractible*.

Example 3.66   Let S^1 (respectively A) denote the circle (respectively the annulus) defined in polar coordinates by $r = 1$ (respectively $1 \leq r \leq 2$). Then the inclusion $i: S^1 \hookrightarrow A$ is a homotopy equivalence.

Figure 3.1 A circle and an annulus

To see this, define a map $\rho: A \to S^1$ in polar coordinates by $\rho(r, \theta) = (1, \theta)$. It is clear that the composite $\rho \circ i$ is equal to the identity map on S^1. On the other hand, we can see that the composite $i \circ \rho$ is homotopic to the identity via the homotopy given in polar coordinates by

$$H(r, \theta, t) = (r - t(r - 1), \theta).$$

The arrows in the figure below show the trajectories of points in the annulus as t increases from 0 to 1.

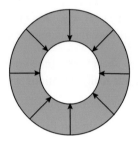

Example 3.67   The sphere in $\mathbb{R}^2$ and the punctured plane $\mathbb{R}^2 - \{0\}$ are homotopy equivalent but not homeomorphic.

Example 3.68   The letters "D" and "P" are homotopy equivalent. They are not homeomorphic.

It is much more difficult to show that two spaces are not homotopy equivalent than to show that they are not homeomorphic. We will be constructing tools for doing this, but in the meantime we will give some examples of spaces which are not homotopy equivalent.

Example 3.69   Intuitively, the "number of holes" in a region in the plane is not changed under homotopy equivalence. Therefore the two spaces in the next figure are not homotopy equivalent.

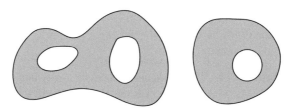

Example 3.70   The circle S^1 and the sphere $S^2 = \{(x, y, z) \mid x^2 + y^2 + z^2 = 1\}$ are not homotopy equivalent. If we view them both as lying in $\mathbb{R}^3$, the sphere has a two-dimensional hole, or "void", and the circle does not.

Example 3.71   If two spaces are homotopy equivalent then they must break up into the same number of connected pieces. It follows that spaces with different numbers of connected pieces are not homotopy equivalent.

3.2.5   Equivalence Relations

This subsection represents a small technical detour. The material in it will be critical to everything we will be doing in the chapters that follow, including the construction of spaces and the building of machinery to detect homotopy-invariant features.

A (binary) relation on a set X is a subset of $X \times X$. We will often denote relations by $\sim$, and write $x \sim x'$ to indicate that x and x' have a relation.

Definition 3.72   A relation $\sim$ on a set X is an *equivalence relation* if the following three conditions hold.

1. $x \sim x$ for all $x \in X$.
2. $x \sim x'$ if and only if $x' \sim x$.
3. $x \sim x'$ and $x' \sim x''$ implies $x \sim x''$.

By the *equivalence class* of $x \in X$, denoted by $[x]$, we will mean the set

$$\{x' \mid x \sim x'\}.$$

The sets $[x]$ for all $x \in X$ form a a partition of the set X. If $\sim$ is any binary relation on a set X, then by the *equivalence relation generated by* $\sim$ (or the *transitive closure of* $\sim$) we will mean the equivalence relation $\sim'$ defined by the condition that $x_0 \sim' x_1$ if and only if there is a positive integer n and a sequence of elements $x'_0, x'_1, \ldots, x'_n$ such that $x'_0 = x_0$, $x'_n = x_1$, and $x'_i \sim x'_{i+1}$ for all $0 \leq i \leq n - 1$.

For a set X and an equivalence relation $\sim$ on X, we will denote the set of equivalence classes under $\sim$ by $X/\sim$, and refer to it as the *quotient* of X with respect to $\sim$.

Definition 3.73   Let X be a set, and let $\sim$ be a binary relation on X. By the *equivalence relation generated by* $\sim$ we will mean the relation $\simeq$, where $x \simeq x'$ if and only if there is a finite sequence $\{x_0, x_1, \ldots, x_n\}$ with $x = x_0$ and $x' = x_n$, and where for each i at least one of the pairs (x_i, x_{i+1}) and (x_{i+1}, x_i) lies in $\sim$. It is readily checked that $\simeq$ is an equivalence relation.

Example 3.74   The rational numbers $\mathbb{Q}$ are typically described as pairs (m, n) of integers, with $n \neq 0$, where m is the numerator and n is the denominator. As we know, the representation is not unique since the pairs (m, n) and (mk, nk) represent the same rational number. We will denote the integers by $\mathbb{Z}$ and the set $\mathbb{Z} - \{0\}$ by $\mathbb{Z}^*$. On the set $\mathbb{Z} \times \mathbb{Z}^*$ we define a binary relation $\sim$ by the requirement that $(m, n) \sim (m', n')$ if and only if there exists a non-zero integer k so that $mk = m'$ and $nk = n'$. We may consider the equivalence relation $\simeq$ generated by $\sim$, and it is easy to check that $\mathbb{Q}$ can be identified with the quotient $\mathbb{Z} \times \mathbb{Z}^* / \simeq$.

Example 3.75   Let X be a finite set, and consider the set $Y = X^n$, the set of n-element sequences with entries in X. There is an equivalence relation $\simeq$ on Y in which two vectors $(x_1, \ldots, x_n)$ and $(x'_1, \ldots, x'_n)$ are equivalent exactly if one can be obtained from the other by permuting the coordinates. In this case the set of equivalence classes consists of all multisets of size n with members in X. Often data comes in ordered form in a spreadsheet, but we are interested in studying it without regard to the ordering. Think for example of shopping baskets, where the order in which the checker checks the objects is not of any relevance. Finding features for data sets of this type, where one wants to ignore some aspects of the data, is often a very interesting problem. We will see a version of this in Section 5.2.

Example 3.76   Let X be the interval $[0, 2\pi]$. Let $\simeq$ denote the equivalence relation $\Delta \cup \{(0, 2\pi), (2\pi, 0)\}$, where Δ is the diagonal $= \{(x, x) \mid x \in X\}$. Then the quotient of X by $\simeq$ can be identified with the circle S^1, via the function that sends $t \in [0, 2\pi]$ to $(\cos t, \sin t)$. Note that this function takes 0 and 2π to the same point in S^1.

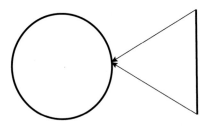

In this example, we have only demonstrated that the circle as a *set* can be identified with the quotient $[0, 1]/ \simeq$. It turns out that if we are given a topological space X and an equivalence relation $\simeq$ on it, there is a naturally defined topology on the set $X/ \simeq$.

Definition 3.77   Let X be a topological space, and let $\simeq$ be an equivalence relation on the set X. We define the *quotient topology* on $X/ \simeq$ to consist of all sets $U \subseteq X/ \simeq$ for which $\pi^{-1}(U)$ is an open set in X, where $\pi : X \rightarrow X/ \simeq$ is the projection given by $\pi(x) = [x]$. It is easy to check that this requirement defines a topology on $X/ \simeq$.

Example 3.76 is one of the simplest cases of this construction. Here is a class of such relations.

Example 3.78   An action of a group G on a set X is a rule that assigns a new element $gx \in X$ to each $g \in G, x \in X$, such that $(gh)x = g(hx)$. If G acts on a set X then the action defines an equivalence relation $\sim_G$ on X by $x \sim_G x'$ if there is a $g \in G$ such that $gx = x'$. This is readily seen to be an equivalence relation, and the equivalence classes are called the *orbits* of the action. If we have an action of a group G on a the underlying set of a topological space, we say the action is *continuous* if, for each element $g \in G$, the function $g \cdot : X \to X$ is a homeomorphism. (Here the multiplication point after g indicates that we are discussing a family of functions g.) In this case, we obtain an equivalence relation on the topological space, and we can form an orbit space using the quotient topology. We will use these ideas extensively in Section 3.2.6, where we generate new spaces from old.

Example 3.79   Let X and Y be topological spaces. We consider the set $F(X, Y)$ of all continuous maps from X to Y, and homotopy constitutes a binary relation on $F(X, Y)$. It is in fact an equivalence relation of $F(X, Y)$, as the following figures show.

The left-hand figure shows that the transitivity property 3 in Definition 3.72 holds, by showing that if f and g are homotopic via a homotopy H and g and h are homotopic via a homotopy H', then f and h are homotopic since it is possible to "splice" H and H' together. The right hand figure demonstrates the symmetry property 2 in Definition 3.72 by showing that a homotopy from f to g can be used to generate a homotopy from g to f by reparametrizing t via the transformation $t \to 1 - t$.

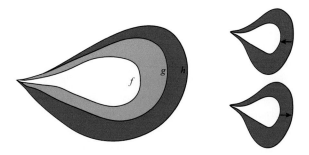

We will denote the set of equivalence classes of maps $f : X \to Y$ by $[X, Y]$.

Up to this point we have dealt with equivalence relations on discrete sets and also on topological spaces. Because much of the work we will be doing to provide invariants of topological spaces relies heavily on algebraic methods, specifically the algebra of vector spaces over a field, we will also introduce the notion of the quotient of a vector space by a subspace. Let V be a vector space over a field $\Bbbk$, and let $W \subseteq V$ be a subspace. We define an equivalence relation $\sim_W$ on V by setting $v \sim_W v'$ if and only if $v - v' \in W$. It is easy to verify that $\sim_W$ is an equivalence relation. We can form the quotient $V / \sim_W$, and one observes that $V / \sim_W$ is itself a vector space over $\Bbbk$, with addition and scalar multiplication rules satisfying $[v] + [v'] = [v + v']$ and $\kappa[v] = [\kappa v]$. In this special case, we will denote $V/\sim_W$ by V/W, and refer to it as the quotient space of V by W. The equivalence classes are referred to as *cosets* or *W-cosets*, and are denoted by $v + W$, where $v \in V$. The element $v \in V$ is clearly not uniquely determined, but any two choices differ only by an element of the subspace W. Although the quotient is an apparently abstract concept, it can be described explicitly in a couple of ways.

Proposition 3.80   *Suppose that we have a basis B of a vector space V and that $B' \subseteq B$ is a subset. If W is the subspace of V spanned by B', then the quotient V/W has the elements $\{[b] \mid b \notin B'\}$ as a basis, so the dimension of V/W is $\#(B) - \#(B')$. More generally, if W' is the complement to W in V, so that $W + W' = V$, and $W \cap W' = \{0\}$, then the composite*

$$W' \hookrightarrow V \xrightarrow{p} V/W$$

is a bijective linear transformation, so that the dimension of V/W is equal to the dimension of W', where p is the map which assigns to $v \in V$ its equivalence class $[v]$ under $\sim_W$.

There is also a matrix interpretation of quotients. Let V and W be vector spaces with ordered bases, and let $f: V \to W$ be a linear transformation, with a matrix $A(f)$ associated with the given bases. The *image* im(f) of f is a subspace of W, and we write $\theta(f)$ for the quotient space $W/\text{im}(f)$.

Proposition 3.81   *Let $g: V \to V$ and $h: W \to W$ be invertible linear transformations. Then $\theta(f)$ is isomorphic to $\theta(hfg)$. It follows that, given the matrix equation $A(f') = A(h)A(f)A(g)$, then $\theta(f')$ is isomorphic to $\theta(hfg)$.*

Proof   This follows from the elementary observation that $w \sim_{\text{im}(f)} w'$ if and only if

$$h(w) \sim_{\text{im}(hf)} h(w').$$ $\square$

Proposition 3.82   *Let W be the vector space $\Bbbk^m$ for some n, and suppose that we are given an $m \times n$ matrix A with entries in a field $\Bbbk$. This matrix can be regarded as a linear transformation from $V = \Bbbk^n$ to W, and the span of its columns is the image of the transformation A. Then if we apply any row or column operation (permuting rows or columns, multiplying a row or column by a non-zero element of $\Bbbk$, or adding a multiple of one row or column to another) to obtain a matrix A', then $\theta(A)$ is isomorphic to $\theta(A')$.*

Remark 3.83   Note that, for any matrix A over a field, one can apply row and column operations to bring it to the form

$$\begin{bmatrix} I_n & 0 \\ 0 & 0 \end{bmatrix},$$

where n is the rank of A. In this case, the dimension of the quotient is readily computed using Proposition 3.80.

We will often want to construct bases for quotients. There is a simple procedure for this process, referred to as *Gaussian elimination*, which puts a matrix into *reduced row echelon form*. It is most commonly described as a method for constructing null spaces of a matrices, but it can also be used to construct bases for quotient spaces. Recall that the standard use of the method proceeds by applying row operations to the matrix until it has the form

$$
\begin{bmatrix}
1 & * & 0 & * & * & 0 & * & * & * \\
0 & 0 & 1 & * & * & 0 & * & * & * \\
0 & 0 & 0 & 0 & 0 & 1 & * & * & * \\
0 & 0 & 0 & 0 & 0 & 0 & 0 & 0 & 0
\end{bmatrix},
$$

where the symbol $*$ denotes an arbitrary element of the field. This is called the reduced row echelon form of the matrix. Formally, it is defined as follows. Given an $m \times n$ matrix A of rank r, we say that it is in reduced row echelon form if it has the following properties.

1. If $i > r$, then the ith row is identically zero.
2. There is a strictly increasing function $\rho \colon \{1, \ldots, r\} \to \{1, \ldots, n\}$ such that (a) $a_{ij} = 0$ for $i \leq r$ and $j < \rho(i)$ and (b) $a_{i,\rho(i)} = 1$ for $i \leq r$. The columns corresponding to the values $\rho(i)$ are called *pivot columns*, and the numbers $\rho(i)$ themselves are called *pivots*. Indices j which are not pivots are called *non-pivots*.
3. $a_{i',\rho(i)} = 0$ for $i' \neq i$.

For a matrix A in reduced row echelon form, there is an explicit description of a basis $\mathcal{B}$ for the null space of A as follows. This basis $\mathcal{B}$ is in bijective correspondence with the set of non-pivot columns of A. For $j \in \{1, \ldots, n\}$ a non-pivot, we construct an n-vector $\vec{v}(j)$ in the null space via the following requirements:

- $\vec{v}(j)_{\rho}(i) = -a_{ij}$;
- $\vec{v}(j)_j = 1$;
- $\vec{v}(j)_{j'} = 0$ for j' a non-pivot $\neq j$.

The vectors $\vec{v}(j)$ for all non-pivots j form a null space for the null space of A.

There is an analogous method for constructing bases of quotient spaces. First, there is evidently a notion of a matrix A in reduced column echelon form, one formulation of which is that A is in reduced column echelon form if and only if the transpose matrix A^T is in reduced row echelon form, and this can be achieved by suitable application of column operations. We will denote the column space of A by $C(A)$. If an $m \times n$ matrix A of rank r is in reduced column echelon form, then there is a function $\rho \colon \{1, \ldots, r\} \to \{1, \ldots, m\}$ whose properties are entirely analogous to the function ρ discussed the in the row echelon form situation. There is also the corresponding notion of a pivot row, and there are also notions of pivots and non-pivots. One can verify that the quotient space of $\Bbbk^m$ by the column space of A admits a basis consisting of the cosets $\vec{e}_i + C(A)$ of the standard basis vectors e_i attached to non-pivots i. It is also important to be able to determine the expansion in terms of this basis of the coset $\vec{v} + C(A)$, where $\vec{v}$ is an arbitrary vector in $\Bbbk^m$; there is an explicit determination of this expansion as follows. First, we observe that it will suffice to obtain the basis expansions of the elements $\vec{e}_i + C(A)$, where i is a pivot, since a general vector can then be obtained as a linear combinations of these elements. For a pivot i, with $i = \rho(j)$, the expansion of $\vec{e}_i + C(A)$ is the linear combination

$$
- \sum_{i' \text{ a non-pivot}} a_{i'j} \vec{e}_{i'}.
$$

This provides an algorithm for determining whether two cosets are the same. It also produces a way to construct the matrix of a linear transformation from a vector space V to a quotient space W/W'.

3.2.6   Constructing Topological Spaces and Maps using Quotients and Products

Modeling data is one of the most important functions of data analysis. In order to do that, we have to have good methods for building models. Since we are performing *topological* data modeling, it is important that we have good ways of constructing topological models, i.e. topological spaces. The formal definition of a topological space requires the specification of an infinite family of sets, namely the open sets. Doing so directly is not a workable option in general, so we must develop methods for specifying topological spaces that are much more compact and which are closer to our geometric intuition. We will also want to perform constructions on spaces that we already understand, in order to construct new ones, and we will want to be able to specify continuous maps. We will introduce the quotient of a topological space, by an equivalence relation, and also the geometric realization of a simplicial complex.

    In addition to constructing spaces from other information, such as a metric, it is also useful to be able to construct new spaces from old. One very useful method is the method of *identification spaces*. The idea is very simple and is nicely exemplified by the well-known example of the Möbius band, which can actually be realized physically. One takes a rectangular strip of paper, and then glues the two ends together with a twist. The gluing is to be thought of as the identification of the two ends with each other. See the following figure.

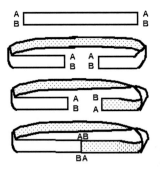

    We can formalize this procedure as follows. We begin with a rectangle $\mathcal{D}$ in $\mathbb{R}^2$, say $[0, 10] \times [-1, 1]$. We will consider the relation $\mathcal{R} \subseteq \mathcal{D} \times \mathcal{D}$ defined by

$$\mathcal{R} = \Delta \cup \{((0, t), (10, -t)), t \in [-1, 1]\}.$$

    The relation $\mathcal{R}$ is an equivalence relation, and we can form the quotient of $\mathcal{D}$ by $\mathcal{R}$. The effect of this is to identify or glue points with x-coordinate equal to 0 with points with x-coordinate equal to 10. The general idea of constructing new spaces from a space X by forming quotients by equivalence relations on X is very useful and we will discuss its properties in this section.

    Recall the notions of equivalence relations, equivalence classes, and quotient topologies from Section 3.2.5.

Example 3.84 (Circle)   Let X denote the space $[0, 1]$, and let $\simeq$ be the equivalence relation whose equivalence classes are of the form $[r] = \{r\}$ when $r \in (0, 1)$ or $\{0, 1\}$. In the quotient, of X by the equivalence relation this means we have identified the point 0 and the point 1. We imagine taking 0 and pulling it to the point 1 and then gluing them together. Intuitively, it is clear that the quotient set is in bijective correspondence with the circle. To show that $X/\simeq$ is actually homeomorphic to the circle requires a pair of results concerning the quotient construction.

Proposition 3.85   *Let $f: X \to Y$ be a map of topological spaces, and suppose that we are given an equivalence relation $\simeq$ on X. Suppose further that whenever $x \simeq x'$, we have that $f(x) = f(x')$. Then there is a continuous map $f_\simeq: X/\simeq \to Y$ which has the property that $f_\simeq([x]) = f(x)$ for all $x \in X$.*

What this allows us to do in the situation of Example 3.84 above is to conclude that, since the continuous map $\varphi: [0, 1] \to S^1$ defined by

$$\varphi(t) = (\cos 2\pi t, \sin 2\pi t)$$

has the property that $\varphi(0) = \varphi(1)$, it factors through a map $[0, 1]/\simeq \to S^1$. It is easy to check that the map is bijective on points, but we would like to conclude that the map is a homeomorphism. We could do this by direct examination of the inverse map in this case, but there is a general result that allows us to conclude the result without any further ado.

Proposition 3.86   *Let X and Y be closed and bounded subsets of $\mathbb{R}^n$; suppose that $\simeq$ is an equivalence relation on x and that we are given an action of a finite group G on Y. Suppose further that we are given a continuous map $f: X/\simeq \to Y/G$, where Y/G denotes the orbit space of the G-action (see Example 3.78), which is a bijection on points. Then f is a homeomorphism.*

Remark 3.87   This result is a specialization of a much more general result, which is beyond the scope of this volume. It requires the notion of compactness as well as the separation axioms, in particular the Hausdorff property, for topological spaces. See Munkres (1975) for this more general result.

Taken together, Propositions 3.85 and 3.86 now give us a proof that we have a homeomorphism $I/\simeq \to S^1$.

Example 3.88 (Torus)   Let X denote the unit rectangle in the plane $\{(x, y) \mid 0 \leq x, y \leq 1\}$. We define an equivalence relation $\simeq$ on X by giving its equivalence classes. These are

$$\begin{cases} \{(x, y)\} & \text{when } 0 < x, y < 1, \\ \{(x, 0), (x, 1)\} & \text{when } 0 < x < 1, \\ \{(0, y), (1, y)\} & \text{when } 0 < y < 1, \\ \{(0, 0), (1, 0), (0, 1), (1, 1)\} & \text{otherwise.} \end{cases}$$

The quotient space is in this case homeomorphic to the usual torus, which can be given simply as $S^1 \times S^1$ with the product topology or which can be realized as a subspace of $\mathbb{R}^4$ as in Example 3.32. Using this second description, we define a map α first on

$[0, 1] \times [0, 1]$ by setting $\alpha(s, t) = (\cos 2\pi s, \sin 2\pi s, \cos 2\pi t, \sin 2\pi t)$. It is easy to check that it respects the equivalence relation and that it then produces a bijection from the set of equivalence classes under $\simeq$ to the points of $S^1 \times S^1$. The result now follows from Propositions 3.85 and 3.86. It is often useful to picture identification spaces of the square using a picture such as the one below, where the colors indicate sides to be identified:

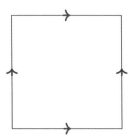

Here the quotient space is a torus.

Example 3.89 (Klein bottle)   Again, let X denote the unit rectangle in the plane $\{(x, y) \mid 0 \leq x, y \leq 1\}$. We define a second equivalence relation $\simeq$ on X whose equivalence classes are

$$\begin{cases} \{(x, y)\} & \text{when } 0 < x, y < 1, \\ \{(x, 0), (x, 1)\} & \text{when } 0 < x < 1, \\ \{(0, y), (1, 1 - y)\} & \text{when } 0 < y < 1, \\ \{(0, 0), (1, 0), (0, 1), (1, 1)\} & \text{otherwise.} \end{cases}$$

The quotient space is in this case is called a "Klein bottle" and is often depicted as in the diagram below.

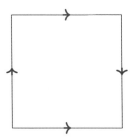

Example 3.90 (Klein bottle, version 2)   Consider the torus $\mathbb{T} = S^1 \times S^1$. As in Example 3.32, we will give it coordinates as the subset of $\mathbb{R}^4$ defined by the equations

$$\begin{cases} x^2 + y^2 = 1, \\ z^2 + w^2 = 1. \end{cases}$$

We define an action of the group $C_2 = \mathbb{Z}/2\mathbb{Z} = \{e, \tau\}$ on $\mathbb{T}$ by

$$\tau(x, y, z, w) = (-x, -y, z, -w)$$

and define the Klein bottle to be the orbit space (see Example 3.78) of the C_2-action, i.e. the equivalence relation consisting of $\Delta \cup \{(v, \tau(v))\}$ for $v \in \mathbb{T}$.

We now define a map from the construction in Example 3.89 to this model. In order to do this, we first define a map β from $[0, 1] \times [0, 1]$ to $\mathbb{T}$ via the formula

$$\beta(s, t) = (\cos \pi s, \sin \pi s, \cos 2\pi t, \sin 2\pi t).$$

As the Klein bottle is the orbit space of a finite group action on a closed and bounded subset of $\mathbb{R}^4$, Proposition 3.86 tells us that if we can prove that the map β respects the equivalence relation and is bijective on points, then we will have proved that it is a homeomorphism. Both of these facts are easy to check.

Example 3.91 (Projective plane)   Let X denote the unit disc in the plane, and let $\simeq$ denote the equivalence class whose equivalence classes are as follows:

$$\begin{cases} \{(x, y)\} & \text{when } x^2 + y^2 < 1, \\ \{(x, y), (-x, -y)\} & \text{when } x^2 + y^2 = 1. \end{cases}$$

The resulting quotient space is called the *real projective plane*, and its points can be identified with the set of all lines through the origin in Euclidean three-dimensional space as follows. First, we can identify the unit disc in the (x, y)-plane with the "northern hemisphere" in the sphere in 3-space by sending each point in the disc to the point on the sphere which lies over it. Next, any line through the origin intersects the northern hemisphere. If the line does not lie in the (x, y)-plane, it will intersect the northern hemisphere in exactly one point. If it lies in the (x, y)-plane, then it intersects the northern hemisphere in two points, which are antipodal to each other. This means that the set of lines through the origin is exactly parametrized by the quotient space described above. A line not in the (x, y)-plane corresponds to the equivalence class containing the unique point of intersection, and a line in the (x, y)-plane corresponds to the equivalence class $\{\alpha, -\alpha\}$, where α is one of the intersection points of the line with the "equator".

Example 3.92 (Cone)   Let X be any topological space. We may then consider the product of X with the unit interval $[0, 1]$, $X \times [0, 1]$. We define an equivalence relation $\simeq$ on this space having equivalence classes

$$\begin{cases} \{(x, t)\} & \text{when } t < 1, \\ \{(x, 1)\}_{x \in X}. \end{cases}$$

The quotient space $X \times [0, 1]/\simeq$ is referred to as the *cone* on X. The point corresponding to the equivalence class $\{(x, 1)\}_{x \in X}$ is called the *cone point*. See the following picture.

Example 3.93 (Suspension)   Again, let X be any topological space. We consider the product of X with the unit interval $[0, 1]$, $X \times [0, 1]$, and define another equivalence relation $\sim$ on this space through $(x, 0) \sim (y, 0)$ and $(x, 1) \sim (y, 1)$ for any $x, y \in X$, leaving each (x, t) equivalent to no other point.

The quotient space $SX = X \times [0, 1]/ \sim$ is referred to as the *suspension* on X, since it can be pictured as X suspended between two cone points. See the following picture, the whole of which shows the suspension on X (X is the red shape).

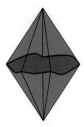

Remark 3.94   These two constructions are homeomorphic to the corresponding constructions made in Section 3.2.3, when the space X in question is a subset of Euclidean space.

There is also the notion of the product of two topological spaces. It is turns out to be useful and corresponds to familiar examples such as $\mathbb{R}^2 = \mathbb{R} \times \mathbb{R}$. Given a pair of sets X and Y, we will denote by $X \times Y$ the set of ordered pairs (x, y), where $x \in X$ and $y \in Y$. The construction $X \times Y$ is called the product of the sets X and Y.

Definition 3.95   Let X and Y denote two topological spaces. Then the product space of X and Y, denoted by $X \times Y$, has the set $X \times Y$ as its underlying set, and a subset of $U \subseteq X \times Y$ is open if and only if for every point $(x_0, y_0) \in U$ there are open sets $U_X \subseteq X$ and $U_Y \subseteq Y$ such that (a) $(x_0, y_0) \in U_X \times U_Y$ and (b) $U_X \times U_Y \subseteq U$.

One construction that combines aspects of the product with aspects of the suspension is the *join* of two spaces.

Example 3.96   Let X and Y denote two topological spaces. Then the join space of X and Y, denoted by $X * Y$, is the quotient of $X \times Y \times [0, 1]$ by the equivalence relation generated by

1. $(a, b_1, 0) \sim (a, b_2, 0)$,
2. $(a_1, b, 1) \sim (a_2, b, 1)$.

The quotient makes any variation in the B factor irrelevant at the point 0 and any variation in the A factor irrelevant at the point 1: $A \times B \times \{0\}/\sim = A$ and $A \times B \times \{1\}/\sim = B$. Intuitively, we can think of the join space as a topological space built out of a disjoint union by adding line segments from each point of A to each point of B.

Many constructions are special cases of the join space. For instance:

- The cone from Example 3.92 is the join space $CX = \{*\} * X$ joining the base space of the cone with the one-point space.
- The suspension from Example 3.93 is the join space $SX = S^0 * X$.
- The join of two spheres is another sphere: $S^n * S^m = S^{n+m+1}$.

3.2.7   Simplicial Complexes

In this subsection we will introduce a method for constructing topological spaces from purely combinatorial information, with no notion of distance or "nearness". This construction will be critical for everything that follows, including in particular the study of point clouds. To give an initial idea about this, we note that the sphere and the boundary of a tetrahedron are homeomorphic. The boundary of the tetrahedron is a union of four triangles, each pair of which intersect in an edge. There are six edges, and each pair of edges intersects in one of four vertices. The combinatorial information, which lists all the triangles, edges, and vertices together with all the inclusion relations that hold among them, turns out to be sufficient to reconstruct the boundary of the tetrahedron, or rather a space homeomorphic to it. It turns out that most spaces that topologists deal with can be described as homeomorphic to a space built as a union of triangles and higher-dimensional objects called *simplices* and that these spaces can be reconstructed simply from sets of such simplices together with the containment relations between them. Precise definitions now follow.

Let $V = \{v_0, v_1, \ldots, v_n\} \subseteq \mathbb{R}^N$ denote a finite subset. We say that V is in *general position* if it is not contained in any affine subspace of dimension $n - 1$ in $\mathbb{R}^N$.

Remark 3.97 In the case of $\mathbb{R}^2$ and $\mathbb{R}^3$, an affine subspace of dimension 1 (respectively 2) is a line (respectively a plane). There is a definition of affine subspaces for higher-dimensional spaces which generalizes this notion in a precise way. Thus, a triple of points in $\mathbb{R}^3$ is in general position if the points are not collinear, and a quadruple of points is in general position if they are not coplanar.

When a set of points V is in general position, we define the *simplex* spanned by the set V to be the *convex hull* of V, i.e. the set of points which can be expressed as a linear combination

$$r_0 v_0 + r_1 v_1 + \cdots + r_n v_n,$$

with $r_0 + r_1 + \cdots + r_n = 1$ and all coefficients $r_i \geq 0$. In low-dimensional cases, the simplex spanned by a pair of points is the line segment spanned by the points, the simplex spanned by a triple of points is the triangle that they span, and the simplex spanned by a quadruple of points is a tetrahedron. The set of points V is called the set of *vertices* of the simplex spanned by V, and the *dimension* of the simplex spanned by V is $\#(V) - 1$. If we have a subset W of V then the simplex spanned by W is a *face* of the simplex spanned by V. For example, a tetrahedron has four two-dimensional faces, six one-dimensional faces, and four zero-dimensional faces (i.e. vertices). We define a *simplicial complex* to be a space X which can be described as a union of a list $\mathcal{L}$ of simplices with the following two properties.

- If a simplex is included in $\mathcal{L}$ then so is any face of it.
- For any two simplices σ and τ in $\mathcal{L}$, the intersection $\sigma \cap \tau$ is a face of both σ and τ.

A *subcomplex* of a simplicial complex X is a collection of simplices belonging to X that satisfies the conditions above. Each of the diagrams below gives examples of two-dimensional simplicial complexes (a color indicates a particular complex). Each is a collection of triangles which intersect in line segments, and the line segments intersect in vertices.

Associated with any simplicial complex is an entirely combinatorial object called an *abstract simplicial complex*.

Definition 3.98   An abstract simplicial complex is a pair (V, Σ), where V is a finite set called the vertex set and where Σ is a family of non-empty subsets of V such that if $\sigma \in V$ and $\emptyset \neq \tau \subseteq \sigma$ then $\tau \in V$ also. The elements of Σ are called *simplices*. For any simplicial complex X, its associated abstract simplicial complex has the vertex set V of X, and a set of vertices is in Σ if and only if it spans a simplex in X. It is easy to check that this definition gives an abstract simplicial complex. We note that the collection Σ decomposes as a disjoint union $\Sigma = \coprod_{k \geq 0} \Sigma_k$, where Σ_k denotes the elements of cardinality $k + 1$; $k + 1$ is referred to as the *dimension* of the simplex. For any abstract simplicial complex $X = (V_X, \Sigma_X)$, a *subcomplex* of X consists of a subset $U \subseteq V_X$ and a collection Σ_U of subsets of U (a) satisfying $\Sigma_U \subseteq \Sigma_X$ and (b) such that the pair (U, Σ_U) is itself a simplicial complex in that it satisfies the condition above.

Given an abstract simplicial complex (V, Σ), we can produce a geometric simplicial complex from it, embedded into $\mathbb{R}^{|V|}$, by associating with the vertices V the standard basis vectors in $\mathbb{R}^{|V|}$ and with a simplex $\{v_0, \ldots, v_n\}$ the convex hull of the corresponding basis vectors. The union of the simplices is a geometric simplicial complex, called the *geometric realization* of the abstract simplicial complex (V, Σ) and denoted by $|(V, \Sigma)|$. It can be shown that, given a geometric simplicial complex X, it is homeomorphic to the geometric realization of its associated abstract simplicial complex.

Example 3.99   Let $J = \{j_0, \ldots, j_k\}$, and let X be the abstract simplicial complex with vertex set J, where the collection of simplices Σ is the collection of all non-empty subsets of J. The complex X is called the *standard k-simplex on J*, and is denoted by $\Delta[J]$. When $J = \{0, \ldots, k\}$, we simply call it the standard k-simplex and denote it by Δ^k. In the cases $k = 0, 1, 2$, and 3, the geometric realizations of Δ^k are homeomorphic to a single point, an interval, a triangle, and a tetrahedron respectively. For general k, Δ^k is homeomorphic to the convex hull of the set of standard basis vectors in $\mathbb{R}^{k+1}$. Up to homeomorphism, $\Delta[J]$ depends only on the cardinality of J, so $\Delta[J]$ is homeomorphic to $\Delta^{\#(J)}$. We will permit ourselves to abuse the notation and denote by Δ^k both the abstract simplicial complex described above as well as its geometric realization.

Example 3.100   Fixing $k > 0$, we let $\partial \Delta^k$ denote the abstract simplicial complex with vertex set equal to $\{0, \ldots, k\}$, where the collection of simplices Σ is the collection of all proper non-trivial subsets of $\{0, \ldots, k\}$. The geometric realization of $\partial \Delta^k$ is homeomorphic to the boundary of the realization of the standard k-simplex, which in turn is homeomorphic to the $(k - 1)$-sphere.

Example 3.101   Let X be any abstract simplicial complex, with Σ as the collection of simplices. Let $\Sigma[k] \subseteq \Sigma$ consist of all subsets of V_X which have cardinality $\leq k + 1$. The pair (V_X, Σ_X) is itself a simplicial complex and will be referred to as the *k-skeleton* of X. We will denote it $X^{(k)}$.

Example 3.102 Fixing an integer $k \geq 0$, and a subset $J \subseteq \{0, \ldots, k\}$, we define the subcomplex $\Delta[J] \subseteq \Delta^k$ to be the abstract simplicial complex whose vertex set is J and whose simplices consist of all subsets of J. The subcomplex $\Delta[J]$ will be referred to as the *face* of Δ^k corresponding to the subset J.

Given two abstract simplicial complexes $X = (V_X, \Sigma_X)$ and $Y = (V_Y, \Sigma_Y)$, a *map of simplicial complexes* from X to Y is a set map $f \colon V_X \longrightarrow V_Y$ such that, for any simplex $\sigma = \{v_0, \ldots, v_k\}$ of X, the set $\{f(v_0), \ldots, f(v_k)\}$ is a simplex of Y. A map of abstract simplicial complexes induces a continuous map $|f| \colon |X| \to |Y|$. In other words, we can use combinatorial information to construct continuous maps as well as spaces.

Let $X = (V_X, \Sigma_X)$ and $Y = (V_Y, \Sigma_Y)$ be abstract simplicial complexes, and suppose that V_X and V_Y are equipped with total orderings $\leq_X$ and $\leq_Y$. The two total orderings give a partial ordering $\leq_{X \times Y}$ on $X \times Y$. By the *product* of X and Y, we mean the abstract simplicial complex $W = (V_W, \Sigma_W)$ with vertex set $V_W = X \times Y$, so that a subset $\sigma \subseteq V_W$ is a simplex if and only if (a) the restriction of the partial ordering $\leq_{X \times Y}$ is a total ordering and (b) the projections of σ onto the sets V_X and V_Y are simplices of X and Y, respectively.

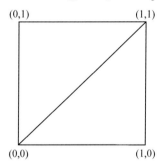

Note that the pair of vertices $\{(0, 1), (1, 0)\}$ is not a simplex because it is not totally ordered, while the pair $\{(0, 0), (1, 1)\}$ is a simplex.

The concept of an abstract simplicial complex is very powerful for creating spaces and maps between spaces. Later, we will see that it can be used to create spaces from finite sets of points sampled from a space X, and that the resulting simplicial complex is often closely related to X. It will also allow us to define the algebraic invariants which we will need to use.

Example 3.103 The torus, seen as an identification space in Example 3.88, can be constructed as an abstract simplicial complex:

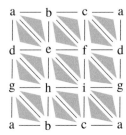

This simplicial complex has nine vertices, 27 edges and 18 triangles. Note that it can be obtained as the product of two simplicial complexes, each isomorphic to $\partial\Delta^2$.

Example 3.104   Just by switching the two vertices d and g along the right edge in Example 3.103, we get the Klein bottle from Example 3.89:

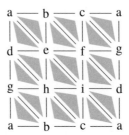

Example 3.105   The OBJ (object file) format for storing geometric meshes for 3D-rendering or 3D-printing is built from single-line instructions:

- a list of v x y z for a vertex at (x, y, z);
- a list of l i j k ... for a line passing through the ith, jth and kth vertices (etc.);
- a list of f i j k for a triangle passing through the ith, jth and kth vertices.

The file format itself allows for polygons that are not strictly triangles. But a triangular mesh without self-intersections, defined in this way, will produce a geometric simplicial complex. Figure 3.2 is one of the most widespread and well-known 3D meshes in the graphics research community and can be acquired as an OBJ file defining a simplicial complex of the surface.

Figure 3.2 Stanford bunny (courtesy of Stanford Computer Graphics Laboratory).

3.2.8   Connectivity Information

The crudest form of qualitative information concerning geometric objects is what we will refer to as connectivity information. It includes parts of all the examples from Section 1.2. What is not included are the corner, edge, and curvature features described in the handprinted letters example there. To discuss these properties, we first define the connected components of a space.

Definition 3.106 For a space X and a point $x \in X$, the **connected component** of x is the set of points in X which can be connected to x by a continuous path in X, i.e. the set of points x' for which there is a continuous map $\varphi \colon [0, 1] \to X$ such that $\varphi(x) = x$ and $\varphi(1) = x'$. In fact, the property of lying in the same connected component is an equivalence relation, because it can be identified with the homotopy relation $*$ on the set of maps from a single point into X, and we saw that this relation is an equivalence relation in Example 3.79. The set of equivalence classes is called the set of connected components of X. We will write $\pi_0(X)$ for the set of connected components of X. Our definition of a connected component corresponds to the term "path component", which is sometimes used by other authors.

The cardinality of the set of components is a numerical *topological invariant* of X in the sense that two spaces which are homeomorphic (or even homotopy equivalent) to each other have the same number of connected components. In the case of the diabetes example from Section 1.2, the indications from the data are that the collection of metabolic and relative weight data of patients suffering from diabetes breaks up into two connected components, one for patients with overt diabetes and the other for patients with the chemical form of the disease.

It turns out that one can view the set of connected components as part of a hierarchy of related invariants. The connected components are regarded as *zeroth-order connectivity information*, and we will now proceed to describe the higher-order versions of it. We will consider the example below of a plane with a disc of radius 1 about the origin removed from it, and provide an informal discussion before making precise definitions.

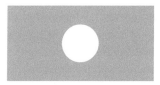

It is easy to convince oneself that this space is connected, since there is a piecewise linear path between any two points in the space:

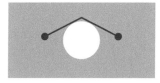

We note, however, that there are many different ways of connecting two points, i.e. many different paths connecting them.

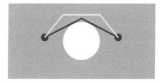

We will choose to view some of these paths as being *essentially the same, or homotopic*. We will say that two paths between a pair of points are essentially the same if there is a continuous family of paths *in the space X* from one to the other which begins with the one path and ends with the other, with the paths in the family always having the same endpoints. Note the similarity between this idea and the idea involved in defining connected components – two points are connected if there is a continuous family of points (i.e. a continuous path) from one of the points to the other. The interesting observation, however, is that some paths in the space X are not essentially the same as some others. For example, in X, the blue and red paths pictured below are essentially distinct.

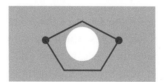

The point is that, try as one might, one cannot drag the red path *continuously* around the hole in X to align it with the blue path while holding the endpoints fixed. One could then say that the goal of first-order connectivity properties is to describe the number of essentially different ways of connecting one point with another one within the same connected component. Unfortunately, one finds that the cardinality of the set of essentially different paths with fixed endpoints in the above situation, while countable and therefore smaller than the set of all paths, is nevertheless infinite. If we consider situations with more holes, the cardinality of the set of essentially different paths is always the same, namely the cardinality of the integers. Fortunately, there is a more sophisticated way of counting that will distinguish these different cases as well as many others. The method requires recognizing the existence of an algebraic structure on some collections of essentially different paths.

Definition 3.107 Let X be a topological space, and let $Y \subseteq X$ be a subspace. Let Z be another space, and let $f : Y \to Z$ be a continuous map. Given two continuous maps $g_0, g_1 : X \to Z$, we say g_0 and g_1 are homotopic relative to Y based at f, if there is a continuous map $H : X \times [0, 1] \to Z$ such that

$$\begin{cases} (x,0) = g_0(x) & \text{for all } x \in X, \\ H(x,1) = g_1(x) & \text{for all } x \in X, \\ H(y,t) = f(y) & \text{for all } y \in Y \text{ and } t \in [0,1] \end{cases}$$

The homotopy relative to Y based at f is an equivalence relation, and we denote its set of equivalence classes by the notation $[(X,Y), Z; f]$

This concept encodes the notions of the homotopy of paths with endpoints fixed at a and b mentioned above, by letting $X = [0, 1]$, $Y = \{0, 1\}$ and $f(0) = a$, $f(1) = b$.

Consider the situation where the endpoints a and b are both equal to $x \in X$, and denote the corresponding set of equivalence classes of paths by $\pi_1(X, x)$. Each map with the given restrictions $g(0) = g(1) = x$ can be thought of as a loop in X, based at x, since the two endpoints of $[0, 1]$ are sent to the same point. We observe that given two loops $g : [0, 1] \to X$ and $g' : [0, 1] \to X$, based at x, we can form a new loop $g * g'$ via the formula

$$\begin{cases} g * g'(t) = g(2t) & \text{for } 0 \le t \le \frac{1}{2}, \\ g * g'(t) = g'(2t - 1) & \text{for } \frac{1}{2} \le t \le 1. \end{cases}$$

Note that this is a loop since $g * g'(0) = g * g'(1) = x$, and the map is continuous since $g(1) = g'(0) = x$. It also turns out that it is possible to make the $*$-operation into a multiplication on the set $\pi_1(X, x)$, that this multiplication is associative, that there is an identity element for the multiplication (the constant path with value x), and that every element $\gamma \in \pi_1(X, x)$ has a unique inverse under the operation $*$. We can summarize this situation by asserting that $\pi_1(X, x)$ is actually a group, rather than just a set. We refer to this group as the *fundamental group* of the space at the base point x. For connected spaces, the choice of base point does not affect the isomorphism class of the fundamental group, so we will allow ourselves not to specify it. It turns out that when we recognize that $\pi_1(X, x)$ is a group, we *are* able to distinguish between the two situations below.

Letting X and Y denote the space on the left and on the right, respectively), we find that $\pi_1(X, x)$ is isomorphic to the free group on one generator, i.e. the additive group of the integers $\mathbb{Z}$. On the other hand $\pi_1(Y, y)$ is isomorphic to the free group on two generators F_2. These groups are clearly not isomorphic (F_2 is non-abelian), so we have managed to distinguish between X and Y. A more careful analysis lets us count the number of holes. Let Z_n denote the space obtained by removing n disjoint discs from the plane. Then the group $\pi_1(Z_n, z)$ is isomorphic to the free group F_n on n generators. It is easy to check that there is an isomorphism invariant called the rank of a finitely generated free group, and that it takes the value n on F_n. It follows that we can count the number of holes by computing the rank on the corresponding fundamental groups.

Remark 3.108 We have implicitly been using the fact that homeomorphic spaces have isomorphic fundamental groups. This is an obvious fact from the definitions. It is also true that homotopy equivalent spaces have isomorphic fundamental groups. This requires a bit of proof (but not much).

For concreteness, it is interesting to develop some intuition about the identification of $\pi_1(X, x)$ with the group of integers $\mathbb{Z}$, for the case of the plane with one hole. It is through the notion of the *winding number*. Specifically, given a loop, one can count the number of times it "winds around the hole", counted with orientation. Loops moving in the counterclockwise direction will be counted as positive, and in the clockwise direction as negative.

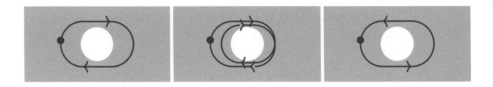

Figure 3.3 Picture with loops with winding numbers n equal to -1, -2, and $+1$

In other words, any loop can be deformed to a model loop which wraps around the hole n times, where negative n corresponds to clockwise movement around the hole and any two loops which have the same *winding number* around the hole can be deformed to each other.

Remark 3.109   Of course, the winding number in the plane can be computed as a line integral over the path, using complex analysis. This approach plays a central role in the robotics case study in Section 6.7.

The fundamental group is to be thought of as *first-order connectivity information*.

A key point of this definition is that it now suggests how to define second-, third-, and higher-order connectivity information.

Let X be a topological space, and let $x \in X$ be a point. Let $I = [0, 1]$. We may construct the n-fold product I^n and the subspace $\partial(I^n) \subseteq I^n$ consisting of all $(t_1, \ldots, t_n)$ such that $t_i = 0$ or 1 for some i. In the case $n = 2$, $\partial(I^2)$ is the boundary of the unit square, and it is in general the boundary of I^n, suitably defined. By an *nth-order loop based at $x \in X$*, we mean a continuous map

$$f : I^n \to X \text{ such that } f(\partial(I^n)) = \{x\}.$$

The set $\pi_n(X, x)$ is now defined to be

$$[(I^n, \partial(I^n)), X; x],$$

where x denotes the constant map with value x. It turns out that $\pi_n(X, x)$ also admits a group structure for $n = 2$. The product $[f] \cdot [g]$ is obtained by placing f and g next to each other to obtain a map $f * g : [0, 2] \times [0, 1] \to X$, and reparametrizing in the t_1 coordinate to obtain a map from $[0, 1] \times [0, 1]$ to X. This procedure can be mimicked for all $n \geq 2$, and gives a group structure on $\pi_n(X, x)$ for all $n \geq 1$. An interesting observation is that π_n is always abelian for $n \geq 2$, as distinct from the case $n = 1$. We give some examples.

Example 3.110   Let Y be the space obtained from $\mathbb{R}^3$ by removing the unit ball centered at the origin; see below.

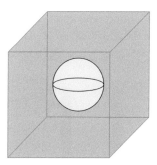

In this case, we have that $\pi_2(Y, y)$ is isomorphic to $\mathbb{Z}$, the additive group of the integers. A generating element is the map of the spheres, which encompasses the hole in the middle. We also have that $\pi_1(Y, y)$ vanishes. Furthermore, the groups $\pi_n(Y, y)$ are not known for all higher values of n. One early discovery was that $\pi_3(Y, y)$ is also isomorphic to $\mathbb{Z}$.

Example 3.111   Let Z be the space obtained from $\mathbb{R}^3$ by removing an infinite cylinder, centered around the z-axis, of radius 1.

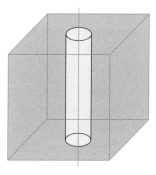

The cylinder we have removed creates a *tunnel* in our space. The presence of the tunnel means that $\pi_1(Z, z)$ is isomorphic to $\mathbb{Z}$, the generating element being a loop running around the tunnel. The space is connected, so $\pi_0(Z)$ consists of a single element, and it turns out that $\pi_n(Z, z)$ is the zero group for all $n \geq 2$.

The homotopy groups were introduced by Čech (1932) and by Hurewicz (1935). As is suggested by Example 3.110 above, the homotopy groups, while conceptually simple and easy to define, are extremely difficult to compute. Even to the current day, the homotopy groups of spheres are only partially understood. In Section 3.3 below, we will introduce a different class of invariants, called homology groups, which are more difficult to define but much simpler to compute. They will be the tools that extend most naturally to data sets.

3.2.9   Hard Features

The connectivity information described above captures much interesting information about geometric objects, but there are many properties that we would regard as qualitative which are not captured directly by topological means. Here are two examples.

Example 3.112   Suppose that we consider the problem of recognizing the difference between a letter U and a letter V. They cannot be distinguished by connectivity information, because they are in fact homeomorphic.

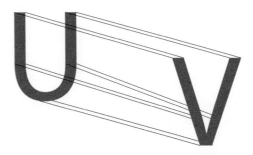

However, the presence of the corner point at the bottom of the letter V, and its absence in the letter U, is something we would still characterize as qualitative. It is something that is retained even if the letters are drawn on a rubber sheet which is stretched, or if the letters are seen from different angles. We would like to be able to recognize such a qualitative property as well as those which are directly detected by connectivity information.

Example 3.113   Consider the problem of distinguishing between a rectangular prism and a tetrahedron. Here, we mean the surfaces of the objects, not the solid objects they bound. They are homeomorphic to each other (in fact, they are both homeomorphic to the two-dimensional sphere).

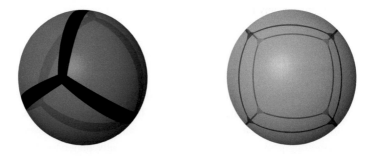

Figure 3.4 Images of homeomorphisms from a prism and a tetrahedron to a sphere

We can distinguish between them intuitively by noting that the rectangular prism has 12 edges and eight vertices, while the tetrahedron has six edges and four vertices. This criterion is also qualitative in nature since it is independent of the lengths of the sides of the figures and since it can be recognized even if the objects are viewed from different angles or even if they are stretched and deformed.

The goal is now to use connectivity information to make such qualitative distinctions. As we have observed, it is not possible to use connectivity information directly since in each case the pair of objects are homeomorphic. The idea will be to construct a new space whose connectivity information does reflect these properties. This space will be called the *tangent complex*, and it is a version of the *tangent cone* construction in geometric measure theory. Here is the definition.

Definition 3.114 Let $X \subseteq \mathbb{R}^n$ be a subset. We define a subset $T^0(X) \subseteq X \times S^{n-1}$ by

$$T^0(X) = \left\{ (x,v) \,\middle|\, \lim_{t \to 0} \frac{d(x+tv, X)}{t} = 0 \right\},$$

where $d(y, X)$ denotes the infimum of the set of numbers $d(y, x)$ as x varies over the set X and d denotes the Euclidean distance; S^{n-1} denotes the unit sphere in $\mathbb{R}^n$. We now define the *tangent complex of X* to be the closure $T(X) = \overline{T^0(X)}$ in $X \times S^{n-1}$.

Remark The point (x, v) lies in $T^0(X)$ if the ray based at X pointing in the direction of the vector v approaches X at a rate that is faster than linear. If the subset X is a smooth submanifold then this yields the usual notion of a tangent vector to X.

Example 3.115 Consider the letter V as a subset of the Euclidean plane. We will embed it as the union of two intervals, namely $[0, 1] \times 0$ and $0 \times [0, 1]$.

Consider any point of the form $\xi = (t, 0)$, with $t \neq 0$. Then (ξ, v) is in $T^0(X)$ if and only if $v = (\pm 1, 0)$. Similarly, if $\xi = (0, t)$ with $t \neq 0$ then $(\xi, v) \in T^0(X)$ if and only if $v = (0, \pm 1)$. In other words, there are only two choices for the v-component of the tangent vector. Next, suppose $\xi = (0, 0)$. Then we can observe that the only choices for v for which $(\xi, v) \in T^0(X)$ are $(1, 0)$ and $(0, 1)$. The following figures illustrate tangents at different parts of the figure near the start of this example. The corner point in the rightmost figure is an exception – here there are two tangent directions, as indicated by the red and green arrows.

To study $T(X)$, we need to take the closure of the collection of these vectors. It is now the case that all four vectors of the form $(\pm 1, 0)$ and $(0, \pm 1)$ are in $T(X)$. To see this notice that any point of the form $((\epsilon, 0), -1)$ is in $T^0(X)$ and that, as we let ϵ tend to zero, we obtain a sequence of points converging to $((0, 0), -1)$. The tangent complex is now pictured as follows.

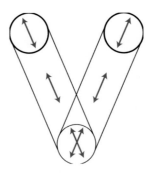

Though we do not see it in this picture, the two red components, just like the two blue components, are actually disjoint from each other. They are on different components of the L-shaped tube formed by the space $X \times S^1$. Note that this space, the tangent complex of X, breaks into *four* connected components. On the other hand, the tangent complex of any smooth connected curve in the plane, such as the letter U, breaks into exactly *two* components, given at each point by the two possible tangent directions available at that point.

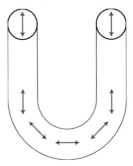

In other words, the connectivity information about the tangent complex of the letters U and V distinguishes them even though the letters themselves are topologically indistinguishable.

This kind of method can be extended to higher dimensions, to successfully distinguish between various polyhedra as well as between other two-dimensional geometric objects in 3-space such as cones, cylinders, spheres, etc. We will explore these possibilities in detail later in the book.

3.2.10   Soft Features

There are other kinds of qualitative information which human beings can recognize and which do not involve singular points, as the example in the previous subsection did. For example, if we consider the problem of recognizing the difference between a printed letter "I" and a printed letter "C", we see that not only are the underlying spaces homeomorphic, but there are no singular points in either figure to distinguish them from each other. Similarly, if we attempt to distinguish a circle from an ellipse then we again see that the spaces are homeomorphic and that there are no singular points to distinguish one from the other. Nevertheless, we regard such distinctions as essentially qualitative.

We are not asking to compare a circle with a given ellipse, but, rather, we are asking how do we distinguish a circle from all ellipses. Similarly, the letter "C" can involve many different values of curvature, but human beings can identify them all as forms of a single prototype letter. We are not interested in maintaining a database of all occurrences of the given object (an ellipse or the letter "C"), but rather in identifying characterizing properties for them. The method we will describe for doing so will be presented only in the case where the underlying geometric object is a smooth hypersurface of $\mathbb{R}^n$. The term "hypersurface" means that the dimension of X is $n-1$, one less than the dimension of the ambient space $\mathbb{R}^n$. The method can be extended to non-smooth situations.

Suppose that $X \subseteq \mathbb{R}^n$ is a smoothly embedded hypersurface. The tangent complex to X can now be given a filtration as follows. Since X is a hypersurface, there is a well-defined perpendicular direction to the tangent plane to X, called the *normal*. For any pair $(x, v) \in T(X)$, the tangent vector v and the normal direction determine a plane in space containing the point x. The intersection of this plane with the surface itself is a smooth curve C in the plane, also containing the point x. For any such C, one may define its *osculating circle*, i.e. the circle containing x which approaches C with "second order contact": not only is the circle tangent to C, its curvature also matches that of C. If the curve C is flat then the osculating circle is actually a line, regarded as a degenerate case of a circle. We will write $\rho(x, v)$ for the radius of the osculating circle (which can take the value $+\infty$), and set

$$\kappa(x, v) = \frac{1}{\rho(x, v)}.$$

For any given real value δ, we now write $T_\delta(X)$ for the subspace of $T(X)$ of pairs (x, v) for which $\kappa(x, v) \leq \delta$. Of course, if $\delta \leq \delta'$ then $T_\delta(X) \subseteq T_{\delta'}(X)$. The qualitative properties distinguishing circles from ellipses can be formulated in terms of "parametrized connectivity information", as the following example shows.

Example 3.116   We will consider a circle S of radius R and a generic ellipse E, and study the connectivity information of the spaces $T_\delta(S)$ and $T_\delta(E)$ to make a distinction between them. We consider S first. In this case, given any point (x, v) in the tangent complex, the corresponding osculating circle is S itself, and therefore $\kappa(x, v) \equiv \frac{1}{R}$. This means that $T_\delta(S) = \emptyset$ for $\delta < \frac{1}{R}$, and that $T_\delta(X) = T(X)$ for $\delta \geq \frac{1}{R}$. The space $T(X)$ can be thought of as being made up of two circles, the first consisting of pairs (x, v) with v pointing in the clockwise direction and the second consisting of pairs where v points counterclockwise. So, $T(X)$ is made up of two connected components, both of which are "born" (i.e. appear) at $\delta = \frac{1}{R}$.

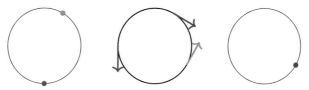

For the ellipse E, on the other hand, we have the following situation. The function κ is no longer constant, but rather achieves its minimum at tangent vectors based at the intersection points of the semiminor axis with E, and its maximum at the corresponding corresponding intersection points with the semimajor axis.

The blue regions correspond to smaller values of κ and the red regions to larger values. We let κ_- and κ_+ denote the minimum and maximum values of κ, respectively. What we now find is first that $T_\delta(X) = \emptyset$ for $\delta < \kappa_-$ and $T_\delta(X) = T(X)$ for $\delta \geq \kappa_+$. For $\delta = \kappa_-$, we find that $T_\delta(X)$ consists of four distinct points, namely the two different tangent directions at the two distinct points of intersection of the semiminor axis with E.

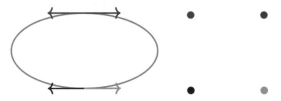

For $\kappa_- \leq \delta < \kappa_+$, $T_\delta(X)$ consists of four components, which form the tangent complex to a union of two disjoint arcs containing the two intersection points of E with the semiminor axis:

So, we see that the set of connected components goes through the following transitions.

$$\emptyset \longrightarrow 4 \text{ components} \longrightarrow 2 \text{ components},$$

with the first transition happening as δ attains the value κ_- and the second one happening as δ attains the value κ_+. Moreover, in the second transition we have two disjoint pairs each merging into a point. For the circle, however, we have the single transition

$$\emptyset \longrightarrow 2 \text{ components},$$

occurring as δ attains the value $\frac{1}{R}$. This kind of transition diagram distinguishes between S and E.

Later in this book we will see how to formalize this into a systematic method for keeping track of such transition diagrams, and how one can obtain similar information about higher-dimensional figures.

3.3 Chain Complexes and Homology

In our discussion of connectivity information in Section 3.2.8, we found that the problem of "counting loops" in a space was best encoded by assigning to the space a group, called the fundamental group. We also constructed higher-dimensional analogues called homotopy groups, which to an extent reflect the presence of higher-dimensional holes in a space. These constructions are conceptually attractive and are the subject of intense study within topology. However, they suffer from some disadvantages.

1. They are extremely difficult to compute. Not even for the two-dimensional sphere are all the homotopy groups known.
2. The connection of higher homotopy with the notion of holes in a space is somewhat tenuous. For example, the three-dimensional homotopy group of the two-dimensional sphere S^2 is isomorphic to the cyclic group $\mathbb{Z}$. Since S^2 is a two-dimensional object, one cannot interpret the three-dimensional homotopy group in terms of holes in it.

In this section we introduce another related family of invariants, called *homology groups*. They are readily computable using linear algebraic methods, and they correspond much better to our notion of counting holes in a space. Their disadvantage is that their definition involves much more algebraic machinery.

3.3.1 Betti Numbers

We will first consider one-dimensional abstract simplicial complexes, which are usually called *graphs*. The idea will be to assign to this combinatorial data a matrix with entries in a field $\Bbbk$, and the invariants in which we are interested will come out of applying Gaussian elimination to this matrix. One convenient choice for $\Bbbk$ is the field $\mathbb{F}_2 = \{0, 1\}$ (see Dummit & Foote 2004), but the fields $\mathbb{F}_p$ for p odd or the rational numbers $\mathbb{Q}$ are also common choices. An advantage of the field $\mathbb{F}_2$ is that it enjoys the property that $1 = -1$, which simplifies the discussion a bit. Additionally, arithmetic in the field $\mathbb{F}_2$ can be thought of as Boolean arithmetic, with addition corresponding to "exclusive or" and multiplication corresponding to "and". To introduce the idea, we will work with the field $\mathbb{F}_2$ and with a particular example, which we can picture as follows:

The vertex set consists of the three-element set $\{0, 1, 2\}$, and the set of one-dimensional faces (called edges) is $\{(01), (02), (12)\}$. There are no two-dimensional simplices.

We create a matrix whose rows (respectively columns) are in one-to-one correspondence with the vertices (respectively edges).

	(01)	(02)	(12)
0	*	*	*
1	*	*	*
2	*	*	*

The entries in the matrix are determined as follows.

- Given an edge e and a vertex v, if $v \notin e$ then the entry in the matrix lying in the column corresponding to e and the row corresponding to v is 0.
- Given an edge e and a vertex v, if $v \in E$ then the entry in the matrix corresponding to this pair is 1.

The matrix for the graph we are considering is

$$\begin{pmatrix} 1 & 1 & 0 \\ 1 & 0 & 1 \\ 0 & 1 & 1 \end{pmatrix}.$$

Remark 3.117   Note that this matrix is intimately connected with the notion of the boundary of an edge. The boundary of an edge consists of the two vertices, and the column of the matrix corresponding to a particular edge is the sum of the rows corresponding to the two members of the of the boundary. For this reason, this matrix will be called the boundary matrix.

Consider the null space of this matrix. It can be determined by performing Gaussian elimination on the matrix on put it into reduced row echelon form. We obtain

$$\begin{pmatrix} 1 & 0 & 1 \\ 0 & 1 & 1 \\ 0 & 0 & 0 \end{pmatrix}.$$

Since there are two pivot columns, it is clear that the null space is one-dimensional and is spanned by the column vector

$$\begin{pmatrix} 1 \\ 1 \\ 1 \end{pmatrix}.$$

If we denote the basis vectors corresponding to the columns by the corresponding edges, we get the vector sum $(01) + (02) + (12)$. Remember that we are working in $\mathbb{F}_2$, so that $1 = -1$. Informally, if we treat each vector in the sum as denoting the corresponding edge in the graph, and if we allow ourselves to interpret sums as unions, then this element has zero (to be interpreted as empty) boundary. The analogy is particularly strong since the sums in $\mathbb{F}_2$ can be interpreted as exclusive or (or as exclusive union).

If we wish to use other fields for the entries of the matrices, we must introduce signs in the boundary matrix. There is a natural way to do this if we select a total ordering on the set of vertices. The matrix entries will now be $+1$ for the smaller vertex in the edge, and -1 for the larger vertex. The matrix for the example above will be

$$\begin{pmatrix} 1 & 1 & 0 \\ -1 & 0 & 1 \\ 0 & -1 & -1 \end{pmatrix}.$$

The analysis of the null space for this matrix is entirely analogous to that for $\mathbb{F}_2$, and the null space is spanned by the vector $(01) - (02) + (12)$. If we interpret the sign in front of (02) as indicating a traversal in the opposite direction, i.e. from 2 to 0, we obtain the interpretation that the loop is being traversed in the counterclockwise direction. The negative of the vector corresponds to traversal in the clockwise direction.

By experimenting with other graphs, one finds that this correspondence indeed reflects our intuition. For example, if we have the graph

then our intuition suggests two loops, and the dimension of the null space of the corresponding matrix is two-dimensional. Similarly, if we take any tree (which can never have a loop), it is not hard to see that the dimension of the null space is zero.

It is of interest to note that the number of connected components of a graph can also be given by a number attached to this matrix, namely the dimension of the quotient space of $\Bbbk^n$ divided by the column space of the boundary matrix, where n is the number of vertices. Suppose we have the example pictured below, whose vertex set is $\{a, b, c, d, e\}$ and whose set of edges is $\{\{a, b\}, \{c, d\}, \{d, e\}, \{c, e\}\}$.

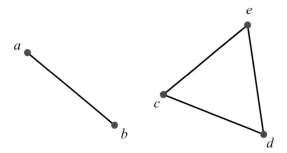

The boundary matrix δ for the field $\mathbb{F}_2$ is

	(ab)	(cd)	(ce)	(de)
a	1	0	0	0
b	1	0	0	0
c	0	1	1	0
d	0	1	0	1
e	0	0	1	1

Its reduced column echelon form is

$$
\begin{pmatrix}
1 & 0 & 0 & 0 \\
1 & 0 & 0 & 0 \\
0 & 1 & 0 & 0 \\
0 & 0 & 1 & 0 \\
0 & 1 & 1 & 0
\end{pmatrix}.
$$

The discussion of reduced column echelon forms and quotient spaces in Section 3.2.5 tells us that a basis for the quotient space is given by the two cosets $b + C(\delta)$ and $e + C(\delta)$, which are representative of the two components of the graph. Note that the cosets $c + C(\delta)$ and $d + C(\delta)$ are both equal to $e + C(\delta)$.

These calculations suggest that it should also be possible to obtain higher-order connectivity information by choosing suitable matrices, obtaining the number of independent higher-dimensional cycles as an appropriate linear algebraic quantity attached to these matrices. This is indeed what happens. For any abstract simplicial complex X, we will define a collection of matrices ∂_k, for each $k \geq 0$, the columns of the matrix ∂_k being in one-to-one correspondence with the k-simplices of X, and its rows being in correspondence with the $(k-1)$-simplices of X. It turns out that the number of independent k-cycles can be interpreted as the following difference: $\mathrm{nullity}(\partial_k) - \mathrm{rank}(\partial_{k+1})$. This number is always non-negative owing to the fact that the matrices ∂_k satisfy the relation $\partial_k \circ \partial_{k+1} = 0$. In summary, the number of independent k-dimensional cycles can be obtained algebraically via linear algebra applied to suitably defined matrices. The number of independent k-dimensional cycles is typically referred to as the *kth Betti number* of the simplicial complex.

3.3.2   Chain Complexes

At the core of this transformation into algebra lies *chain complexes*. These are algebraic representations of simplicial complexes.

Definition 3.118   A *chain complex* $V_* = (V_*, \partial_*)$ is a sequence $\ldots, V_i, V_{i-1}, \ldots$ of vector spaces together with a sequence of linear transformations $V_i \xrightarrow{\partial_i} V_{i-1}$ such that $\partial_i \circ \partial_{i+1} = 0$. The elements of a given V_i are called *i-chains*, and the map ∂_i is the *boundary map* or the *differential*.

We shall often write V_* or just V for the collection $\{V_i\}_{i \in \mathbb{Z}}$ and ∂_* or just ∂ to refer to all the ∂_i. In this way, we may view a chain complex as a single, large vector space $V = \bigoplus_i V_i$, and the differential as a single linear transformation $\partial: V \to V$ such that $\partial^2 = 0$.

In order to define a chain complex attached to an abstract simplicial complex, we will need the concept of the *free vector space on a set*. The exact definition of this construction is somewhat complex, so here we will merely describe its properties, which are sufficient for all the constructions in which we are interested.

Proposition 3.119   *Let* $\Bbbk$ *be a field and* X *a set. Then there is an associated vector space* $F(X)$, *and the assignment* $X \to F(X)$ *has the following properties.*

1. *There is an inclusion of sets $i_X : X \to F(X)$, and the image of i_X is a basis for $F(X)$. In particular, the dimension of $F(X)$ is equal to $\#(X)$ when X is finite.*
2. *For any map of sets $f : X \to Y$, there is a linear transformation $F(f) : F(X) \to F(Y)$ such that the diagram*

$$
\begin{array}{ccc}
X & \xrightarrow{\ f\ } & Y \\
{\scriptstyle i_X}\downarrow & & \downarrow{\scriptstyle i_Y} \\
F(X) & \xrightarrow{\ F(f)\ } & F(Y)
\end{array}
$$

commutes. This means that the composite map of sets $F(f) \circ i_X$ is equal to the map $i_Y \circ f$.
3. $F(id_X) = id_{F(X)}$.
4. *If we have a sequence of maps*

$$X \xrightarrow{\ f\ } Y \xrightarrow{\ g\ } Z$$

then $F(g) \circ F(f) = F(g \circ f)$.

We will refer to $F(X)$ as the free $\Bbbk$-vector space on the set X.

Remark 3.120 In the terminology of category theory, properties 3 and 4 above assert that F is a *functor* from the category of sets to the category of $\Bbbk$-vector spaces. The additional properties 1 and 2 assert that F is a *monad*, also in category-theoretic terminology. See MacLane (1998) and Riehl (2017) for a development of category theory. It is sufficient to think of F as a rule which assigns to a set a vector space with basis X and which assigns linear transformations to set maps in such a way as to preserve the bases. We will return to this concept later in the book.

Definition 3.121 Let k be a field. Given an abstract simplicial complex Σ, we define the *simplicial chain complex with coefficients in* $\Bbbk$ on Σ, denoted by $C_*(\Sigma)$, by letting $C_n(\Sigma)$ equal the free $\Bbbk$-vector space on the set of n-dimensional faces of Σ.
   To define the boundary map we pick an ordering of the vertices of Σ, thus yielding an ordering on the elements of any face of Σ, and then set

$$\partial_n(\{v_0, \ldots, v_n\}) = \sum_{i=0}^{n} (-1)^{i+1} (\{v_0, \ldots, v_n\} \setminus \{v_i\}).$$

We have defined ∂_n on a basis for $C_n(\Sigma)$, and the map extends by linearity to the entirety of $C_n(\Sigma)$ since the set of n-simplices of Σ forms a basis for $C_n(\Sigma)$. If we wish to be clear about the particular field of scalars $\Bbbk$, we write $C_*(\Sigma; \Bbbk)$ to emphasize it.

Remark 3.122 Note that if the field $\Bbbk$ is $\mathbb{F}_2$ then the signs are irrelevant and we are not required to select an ordering on the vertices.

   An important result about the definition of the linear transformations ∂_n is the following.

Proposition 3.123   *For the vector spaces $C_n(\Sigma)$ and operators ∂_n, we have the identity*

$$\partial_{n-1} \circ \partial_n \equiv 0.$$

This means that the data $\{C_n(X), \partial_n\}$ is a chain complex.

Proof   There is a two-parameter sum defining the composite $\partial_{n-1} \circ \partial_n$, and the summands in the sum are parametrized by pairs $\{i, j\} \subseteq \{0, \ldots, n\}$ with $i \neq j$:

$$\partial_{n-1} \circ \partial_n(\{v_0, \ldots, v_n\}) = \sum_i \sum_{j \neq i} \pm \{v_0, \ldots, v_n\} \setminus \{v_i, v_j\}.$$

The set of such pairs is broken into two disjoint subsets, one in which $i < j$ and the other in which $i > j$. Any $(n-2)$-simplex σ_0 contained in $\sigma\{v_0, \ldots, v_n\}$ is obtained by deleting two elements from σ, and this can be done in two ways, as follows. The coefficient of σ_0 obtained in the sum is the sum of two terms, one in which the deletions occur with the smaller element first and the second in which it occurs last. If the two elements being deleted are i and j, with $i < j$, then the first coefficient is $(-1)^{i+1}(-1)^j$. If, on the other hand, $i > j$, then the coefficient is $(-1)^{i+1}(-1)^{j+1}$. These two coefficients clearly cancel, which verifies that the coefficient of σ_0 is $= 0$. $\square$

Now we return to the example from Section 3.3.1, with the picture

We will write e_σ for the basis element belonging to a simplex σ. There are three zero-dimensional simplices, $0 = \{0\}$, $1 = \{1\}$ and $2 = \{2\}$, and three one-dimensional simplices, $(01) = \{0, 1\}$, $(02) = \{0, 2\}$, and $(12) = \{1, 2\}$.

Thus, $C_0(\Sigma) = \Bbbk^3$ and $C_1(\Sigma) = \Bbbk^3$. The boundary map is formed by the mappings $e_{(01)} \mapsto e_1 - e_0$, $e_{(02)} \mapsto e_2 - e_0$, and $e_{(12)} \mapsto e_2 - e_1$. Thus, the boundary turns out to be given by the matrix we derived in Section 3.3.1.

3.3.3   Homology Groups

As we saw in Section 3.3.1, graphs with cycles have boundary matrices with null spaces corresponding to the cycles, while graphs without cycles have boundary maps with trivial null space. This is far from a coincidence; rather, it is the intuition on which the subsequent theory builds.

A cycle in a graph will have vanishing boundary, since the sum of basis vectors corresponding to any two subsequent edges has a boundary where their meeting point is cancelled out,

$$\partial(ab + bc) = \partial(ab) + \partial(bc) = b - a + c - b = c - a,$$

and thus, if a sequence of edges closes on itself, the two endpoints of the path, i.e. the boundaries of the path, will cancel each other out.

With this as underlying intuition, we define the *boundary group* of dimension n to be $B_n(\Sigma) = \text{Im}(\partial_{n+1})$ and the *cycle group* of dimension n to be $Z_n(\Sigma) = \text{Ker}(\partial_n)$. The condition $\partial^2 = 0$ corresponds to the intuition that a cycle of edges corresponds to a 1-chain without boundary and that this generalizes to any dimension. Furthermore, $\partial^2 = 0$ implies directly that $B_n(\Sigma) \subseteq Z_n(\Sigma)$ for all n. Recall that in the discussion of the fundamental group in Section 3.2.8, we were led to require that pairs of paths that are homotopic, i.e. for which there exists a deformation from the one to the other, should be regarded as identical in the group. Similarly, in this situation we are motivated to identify pairs of cycles for which there is a collection of simplices whose boundary is the union of the two. As an example, consider the simplicial complex below. It has four vertices, five edges $e_1 - e_5$, and one triangle, depicted in color.

In the language of boundaries and cycles, we see that in the left-hand part of the complex we have two different cycles that go around the empty space in the right-hand part, namely $e_1 + e_4 + e_5$ and $e_1 + e_2 + e_3 + e_4$, as shown below:

and

The difference between the two paths in the chain complex is precisely $e_2 + e_3 - e_5$, which we may recognize as the boundary of the colored triangle. This calculation motivates us to form the *quotient* of the cycles by the boundaries. In that quotient group, the difference between the cycles is 0, and therefore the cycles represent the same element.

This leads us to the definition of *homology*.

Definition 3.124 The nth *homology group* of a chain complex C_* is the group $H_n = \text{Ker}(\partial_n)/\text{Im}(\partial_{n+1})$.

For an abstract simplicial complex Σ, its *homology with coefficients in a field* $\Bbbk$ is the homology $H_*(\Sigma; \Bbbk) = H_*(C_*(\Sigma, \Bbbk))$ of its corresponding simplicial chain complex. If the scalar field can be understood from the context, we write $H_*(\Sigma)$.

Remark 3.125 It is possible to replace a field by a *ring*, such as the integers $\mathbb{Z}$. When that is done, the classification of the possible groups becomes more complicated in that it does not consist exclusively of a dimension but rather of a rank, which is analogous to the dimension of a vector space, together with various torsion coefficients. In this case, we write $H_i(X, \mathbb{Z})$ for the homology with coefficients in $\mathbb{Z}$.

The element of H_n that corresponds to the equivalence class of a given cycle z is denoted by $[z]$.

3.3.4 Cochains and Cohomology

Duality, by which we mean the study of functions on an object X rather than the object itself, is often a powerful perspective in algebra. This is also true in algebraic topology. By dualizing homology we get *cohomology*, which will drive several applications and constructions later in the book.

For any $\Bbbk$-vector space V, we may consider the set V^* of linear transformations from V to the one-dimensional $\Bbbk$-vector space $\Bbbk$. Since we may add and multiply by scalars such transformations pointwise, it turns out that V^* is also an $\Bbbk$-vector space. This is a very familiar construction and its properties can be found in Dummit & Foote (2004). If $L \colon V \to W$ is a linear transformation, we can define a linear transformation $L^* \colon W^* \to V^*$ by the formula $L^*(\lambda) = \lambda \circ L$. This property, where the direction of the induced map is opposite to the direction of the original map, is referred to as *contrafunctoriality*.

Given a chain complex C_*, we define its *dual cochain complex* C^* to be the family of vector spaces $C^n = C_n^*$ equipped with coboundary operators $\delta^n = \partial_n^*$. Note that $\delta^n \colon C^n \to C^{n+1}$. It is immediate from the corresponding fact for homology that $\delta^{n+1} \circ \delta^n \equiv 0$. By analogy with the construction of a homology, we can define the cohomology of the complex C_* in dimension n to be the quotient

$$\mathrm{Ker}(\delta^n)/\mathrm{Im}(\delta^{n-1}).$$

Given a simplicial complex Σ, we define the cohomology of Σ to be the cohomology of the chain complex $C(\Sigma)$, and we will denote it by $H^n(\Sigma)$. The elements of $C_n(\Sigma)^*$ are referred to as *n-cochains*. One basis of this space is given by the *dual cosimplices*: if $\sigma \in \Sigma$ is a k-simplex then $\hat{\sigma} \colon C_k(\Sigma, \Bbbk) \to \Bbbk$ defined by $\hat{\sigma}(\sigma) = 1$ and $\hat{\sigma}(\tau) = 0$ for all other τ extends to a linear function. The collection of all such linear extensions forms a basis for the entire cochain group.

To build an intuition for how these vector spaces act and interact, it is helpful to focus on just a single dimension.

Thus, 1-cocycles are functions f from edges to $\Bbbk$ such that $\delta f = 0$. This means that $(\delta f) = (f \circ \partial) \colon C^2 \to \Bbbk$ is the zero map. We can write this out in detail for some 2-simplex $[abc]$, where the square brackets again denote an equivalence class; thus $\delta f([abc]) = f([ab] - [ac] + [bc]) = f(ab) - f(ac) + f(bc)$. This is equal to 0 precisely when $f(ab) + f(bc) = f(ac)$, or in other words precisely when f is invariant over the taking of different paths between the same endpoints.

There are some cases when such a path invariance will not come as a surprise to anyone. For example, if g is a function assigning a *potential* to each vertex, we can construct a function on edges that measures the potential change by following that edge. We see that this construction, which creates a function $[a, b] \mapsto g(b) - g(a)$, is really just the coboundary construction on the 0-cochain g, building the function δg to measure potential changes.

Cohomology therefore can be seen to enumerate precisely the equivalence classes of edge functions that achieve path independence without necessarily being the result of an underlying potential field.

Computationally, it is worth observing that, with the choice of bases given by the simplices for $C_k(\Sigma, \Bbbk)$ and the cosimplices for $C^k(\Sigma, \Bbbk)$, the boundary and the coboundary are represented by related matrices: the coboundary matrix is the transpose of the boundary matrix. This observation leads to a significant speeding-up when one is constructing algorithms for computing with homology or cohomology, and especially so when we arrive at their persistent versions.

Remark 3.126 As mentioned in Remark 3.125, a version of homology has coefficients in the ring of integers $\mathbb{Z}$. There is a corresponding notion of cohomology with integer coefficients, denoted by $H^i(X, \mathbb{Z})$.

3.3.5 Kirchhoff's Laws

An electrical circuit can be represented by a graph, with edges corresponding to wires and components, and vertices corresponding to connection points. Each edge may be labeled with the component it contains and, if any, the total resistance along the edge.

Once we introduce a voltage source to the circuit, each edge may be seen as decorated with more data: that is, with the current flowing through the edge and with the voltage differential along the edge. Orienting the edges, the current direction and voltage differential direction are indicated with arrows giving the signs of their values.

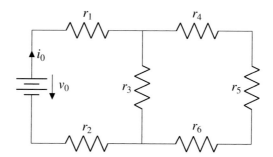

Thus, for an electric circuit, the current flowing through the circuit gives rise to a 1-chain I: a linear combination of edges, with the coefficients for each edge given by the current through that edge relative to its orientation.

The flow of current through each vertex gives rise to a 0-chain, i.e. the assignment of a value to each vertex. Specifically, if we consider a given vertex, the flow through it is the sum of currents for the edges leading out of it, minus the sum of currents for the edges leading into it. See the following diagram.

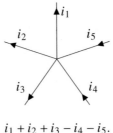

$$i_1 + i_2 + i_3 - i_4 - i_5.$$

Aggregating all these flows into a single 0-chain, each edge contributes its current to exactly two vertices: its source and its target, and with opposite signs. Hence an edge from a vertex s to a vertex t with current i flowing through it will contribute $i(s-t)$ to the complete 0-chain.

From our knowledge of homology, we recognize this as the fundamental formula to compute the *boundary* of a 1-chain and thus we may conclude that the 0-chain of flows is the boundary ∂I of the current 1-chain.

Now, Kirchhoff's first law states that the net flow through each vertex in a closed circuit is 0. With the above analysis, we see that this amounts to the requirement $\partial I = 0$, or in other words that I is a cycle.

We may add cells to the graph: one cell for each cycle in the graph. The result will have vanishing homology in degree 1, since any cycle that could produce a homology will be the boundary of a cell. Since the homology in degree 1 vanishes, each cycle I is a boundary and thus there is some 2-chain J such that $I = \partial J$. These cells are called *meshes* in electromagnetic theory, and, for a mesh M, the resulting 2-chain J bounded by the current chain is called the *mesh current*.

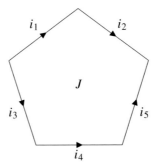

$$J = i_1 + i_2 - i_3 - i_4 - i_5.$$

Adding in all possible meshes, however, overdetermines the situation for the graph under consideration. Certainly, an overdetermined system still models the circuitry but more commonly in mesh analysis we pick a family of (small) representative cycles for the degree-1 homology of the original circuit, and then fill in the *essential meshes* corresponding to this basis; in other words, we pick out *independent cycles* from the graph, leaving the cycles whose edges carry current that will be determined by the chosen meshes, and assign meshes and mesh currents to the independent cycles. By picking precisely one mesh for each homology basis element, we are guaranteed that the result is a contractible space, and thus we may produce even stronger statements than the vanishing of the homology in degree 1, used above.

—————————————— For the advanced reader ——————————————
In fact, the fact that picking out all possible meshes induces higher-degree homology suggests that one might want to choose higher abstractions of meshes to fill in the resulting *cavities*. Thus, we could fill in 3-cells that represent relations between the 2-cells that have been inserted, which, if we insert all possible 3-cells, is bound to produce yet higher-dimensional phenomena. Proceeding in this way indefinitely will produce a circuit represented by an infinitely long chain complex: a *free resolution* of the original circuitry. While the utility of free resolutions to circuit analysis is not transparent, they are important in algebra and geometry to analyze interactions between the equations defining algebraic or geometric objects.

As for the voltage, there is value in considering it to be a 1-cochain instead of a 1-chain. Equally with the current 1-chain, voltage associates a value to each edge but the interactions with boundaries and coboundaries and the laws obeyed by the voltage are different.

Specifically, Kirchhoff's second law states that, for any cycle in the graph, the voltage changes around that cycle have to vanish. Thus, consider a figure like this:

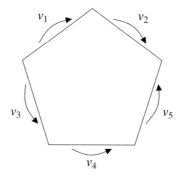

The law says that $v_1 + v_2 - v_3 - v_4 - v_5 = 0$. This, we may recognize, means that the voltage cochain must be a cocycle: around each mesh M, the voltage evaluated at its boundary vanishes.

Just as with the chain complex case for currents, with all meshes in place the degree-1 cohomology vanishes, and thus all cocycles have to be coboundaries. A voltage cocycle V is therefore the coboundary of a quantity ϕ called the *electrostatic potential*; $V = \delta\phi$, having a value at each vertex such that the voltage across an edge is simply the change in potential going along that edge.

Having established currents in closed circuits as 1-boundaries, and voltages as 1-coboundaries, the next step is to relate them to each other. The core of their relations lies in Ohm's law, $V = RI$, where R is the resistance of the edge across which the potential change is V. Since the relationship holds on an edge-by-edge basis, we may consider R as a transformation from 1-chains to 1-cochains. Its edge-by-edge definition means that R is, in fact, a linear transformation, with $V = R(I)$.

The *power* of a given circuit measures the heat dissipation from the circuit. It is given, for any given component, by the formula VI, and thus, in the interpretation of voltage as a cochain and current as a chain, by $V(I)$. So, the power is, using R, given by $V(I) = (RI)(I)$. In physical situations, power is always non-negative, and as long as we are not dealing with vanishing resistances, the only case when the power vanishes is if $I = 0$. Thus, the above formula yields an inner product on the 1-chains, namely, $\langle -, - \rangle$ such that $\langle I, I \rangle = (RI)(I)$.

———————————— For the advanced reader ————————————

Thus the resistance R is not only a transformation, but also an isomorphism between a finite-dimensional vector space C_1 and its dual C^1. When viewed as such an isomorphism, we tend to call R the more general term *impedance*, Z, and its inverse R^{-1} the *admittance* A. This isomorphism induces the isomorphism $H_1 \to H^1$ from Hodge theory.

This example expands to the point where algebraic topology forms a good framework to discuss Maxwell's equations and to formulate large parts of the theory of electromagnetism in a computation-friendly way. For an illustration of the simplicity possible, we may write $\mathbf{F}$ for the combination of the electric and the magnetic field into a 2-form on a 4-dimensional spacetime manifold, $\mathbf{J}$ for the electric current 3-form, $d\omega$ for the exterior derivative, and $*\omega$ for the Hodge star operator that takes a generator to its *complement*, so that $*dx_1 = dx_2 \wedge dx_3 \wedge dx_4$ and $*(dx_1 \wedge dx_3) = dx_2 \wedge dx_4$.

With these definitions, we require the current 3-form to fulfill $d\mathbf{J} = 0$, and Maxwell's equations can be rewritten as

$$d\mathbf{F} = 0, \qquad d * \mathbf{F} = \mathbf{J}.$$

Gross & Kotiuga (2004) put much of electromagnetism on a topological footing, introducing the necessary algebraic topology with a generous supply of examples drawn from electromagnetism. For a much deeper exposition than our space limitations allow, we recommend this as a starting source.

————————————————————

We take the definition of power as a voltage cochain applied to a current chain, $V(I)$. Kirchhoff's laws can be written as the two observations

$$V = \delta P, \qquad \partial I = 0.$$

So, for any circuit obeying Kirchhoff's laws, the *signed* total power is $V(I) = (\delta P)(I) = P(\partial I) = P(0) = 0$. This result is known as Tellegen's theorem (Tellegen 1952) and generates a wide variety of consequences in electrical network theory. The theorem holds in an almost absurd level of generality: not only is there no dependence on what kinds of elements the network contains (linear? non-linear? active? passive? time-varying?) but the voltage cochain and the current chain do not even need to be taken from the same network. As long as the underlying graph is isomorphic, the voltage and current could be taken from networks with different components.

For another exposition of the material here, Baez (2010) has an eminent and inspiring overview as part of a series that compares networks in several situations: electrical, pneumatic, and others.

3.3.6 Chain Maps

A vector space is called *graded* if it can be decomposed into a direct sum $V_* = \bigoplus_{i \in \mathbb{Z}} V_i$. Thus, a chain complex C_* can be viewed as a graded vector space by forgetting about the boundary operator. The summand V_i is called the *(homogenous) component of degree i*, and an element of a single summand $v \in V_i$ is called a *(homogenous) element of degree i*.

A *graded map* $f_*: V_* \to W_*$ of degree d between two graded vector spaces is a collection of linear functions $f_i: V_i \to V_{i+d}$. With this terminology, we can redefine (equivalently) a chain complex to be a graded vector space C_* equipped with a graded map $\partial: C_* \to C_*$ of degree -1 such that $\partial^2 = 0$.

We define a *chain map* $f_*: C_* \to D_*$ between chain complexes C_* and D_* to be a graded map of degree 0 such that $f_* \partial = \partial f_*$, or, in more detail, such that $f_{n-1} \circ \partial_n = \partial_n \circ f_n$ for all n.

Proposition 3.127 *Let $f: C_* \to D_*$ be a chain map. Then f induces a linear transformation $H_n(f): H_n(C_*) \longrightarrow H_n(D_*)$, which has the properties (a) that $H_n(id_{C_*}) = id_{H_n(C_*)}$, (b) that $H_n(f \circ g) = H_n(f) \circ H_n(g)$, and (c) that $H_n(f) = f$ on chain complexes C_* is non-zero in only one dimension i, and therefore $H_i(C_*) \cong C_i$.*

Proof Because f is a chain map, we verify that f restricts to linear transformations $f|_{Z_n(C_*)}$ and $f|_{B_n(C_*)}$. It follows that the definition

$$z + B \mapsto f(z) + B$$

gives the required transformation. □

The following is the most important example for us.

Example 3.128 Let X and Y be two simplicial complexes, equipped with total orderings on their sets of vertices. Let $f: X \to Y$ be a map of simplicial complexes, so that, for every simplex $\sigma = \{x_0, \ldots, x_k\}$ of X, the set map $f|_\sigma: \sigma \to Y$ is weakly order-preserving, in the sense that it preserves the relation $\leq$. Then we obtain an induced chain map $C_*(X) \to C_*(Y)$ by sending each k-simplex of X into its image as a simplex of Y. If a given simplex σ is sent to a simplex of lower cardinality, the chain map is defined to be 0 on σ.

The requirement that the complexes have to be equipped with a vertex ordering, and with the resulting map, is awkward and can be circumvented. The first observation is that if the field is $\mathbb{F}_2$, there is no requirement on orderings and we obtain a chain map in a straightforward fashion. For other fields, where $1 \neq -1$, one can use the notion of the subdivision of an abstract simplicial complex to obtain the definition of an induced linear transformation attached to a map of simplicial complexes. The discussion that follows is somewhat technical and not terribly enlightening, so, for the reader who is willing to consider only the coefficient field $\mathbb{F}_2$, nothing will be lost by skipping it.

Definition 3.129 Let $X = (V_X, \Sigma_X)$ be an abstract simplicial complex. By the *barycentric subdivision of X* we mean the abstract simplicial complex whose vertex set is the set of simplices Σ_X of X and for which a set of simplices $\{\sigma_0, \ldots, \sigma_k\}$ of cardinality k spans a k-simplex if and only if there is a permutation π of the set $\{0, \ldots, k\}$ such that

$$\sigma_{\pi(0)} \subseteq \sigma_{\pi(1)} \subseteq \cdots \subseteq \sigma_{\pi(k)}.$$

We denote this complex by $Sd(X)$. Its set of vertices is equipped with a partial ordering by containment, so $\sigma \leq \tau$ if and only if $\sigma \subseteq \tau$. This partial order admits an extension to a total ordering. We also note that, for any map $f\colon X \to Y$ of abstract simplicial complexes, there is an evident map $Sd(f)\colon Sd(X) \to Sd(Y)$ of abstract simplicial complexes, defined on vertices by $\sigma \to f(\sigma)$, where σ is a simplex of X regarded as a vertex of $Sd(X)$.

Example 3.130   The subdivision of the standard 2-simplex looks like this;

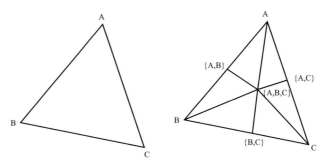

Note that the spaces corresponding to the complexes are homeomorphic. More generally, if X is an abstract simplicial complex, and if we have geometric simplicial complexes Y and Z, whose underlying abstract simplicial complexes are isomorphic to X and $Sd(X)$, respectively, then Y and Z are homeomorphic.

Let X be a simplicial complex equipped with a total ordering on its vertex set. We define a set map $\rho_X\colon V_{Sd(X)} \to V_X$ by $\sigma \to \max(\sigma)$. Note that σ is a set of elements of V_X, and the maximum is computed in the given ordering on V_X. Selecting any total ordering on $V_{Sd(X)}$ which restricts to the canonical partial ordering, the set map ρ_X is a map of simplicial complexes and further satisfies the requirements described in Example 3.128. Therefore we have an induced chain map $C_*(\rho_X)$.

Proposition 3.131   *The induced maps $H_n(C_*(\rho_X))$ are all isomorphisms.*

Proof   A proof can be found in Hatcher (2002).   □

The value of this result is that it permits us to define induced maps corresponding to maps of abstract simplicial complexes using only the subdivided complex, which does not require a choice of ordering. The idea is as follows. Given a map of abstract simplicial complexes $f\colon X \to Y$, we have the corresponding map $Sd(f)\colon Sd(X) \to Sd(Y)$. We have the canonical partial orderings on vertex sets from Example 3.128, and $Sd(f)$ preserves them. If we extend each of them to a total ordering on the vertex sets, the requirements of the two extensions that they restrict to the same total ordering on all simplices are fulfilled. It follows that we have a well-defined linear transformation

$H_n(Sd(f))\colon H_n(Sd(X)) \to H_n(Sd(Y))$. To conclude the discussion, we define the induced map on the homology attached to a map $f\colon X \to Y$ of simplicial complexes to be the composite

$$H_n(X) \xrightarrow{H_n(\rho_X)^{-1}} H_n(Sd(X)) \xrightarrow{H_n(Sd(f))} H_n(Sd(Y)) \xrightarrow{H_n(Sd(\rho_Y))} H_n(Y).$$

The conclusion is the following.

Proposition 3.132 *There is a unique assignment of linear transformations* $H_n(f)\colon H_n(X) \to H_n(Y)$ *to maps* $f\colon X \to Y$ *of abstract simplicial complexes which satisfies the following conditions.*

1. $H_n(id_X) = id_{H_n(X)}$ *for all abstract simplicial complexes* X.
2. $H_n(g \circ f) = H_n(g) \circ H_n(f)$ *for all sequences of maps*

$$X \xrightarrow{f} Y \xrightarrow{g} Z$$

of abstract simplicial complexes.

3.3.7 Chain Homotopies

The idea of a homotopy between two continuous functions has its correspondence in the world of chain complexes as well. We recall that, given two continuous functions between topological spaces $f, g\colon X \to Y$, a homotopy $f \Rightarrow g$ is a continuous function $h\colon I \times X \to Y$ such that $h|_{0 \times X} = f$ and $h|_{1 \times X} = g$. The topological space $I = [0, 1]$ is homeomorphic to the simplicial complex consisting of a single edge x and its two endpoints a, b, so we can consider the corresponding chain complex I_*,

$$0 \to \Bbbk x \xrightarrow{\begin{pmatrix} 1 & -1 \end{pmatrix}} \Bbbk a \oplus \Bbbk b \to 0,$$

with non-zero components in degrees 0 and 1 as a candidate for a model for I in the chain complex world.

Hence, starting with two chain complexes C_* and D_*, we may consider the chain complex $I \otimes C_*$, which in degree n is defined to be $(C_* \otimes I_*)_n = C_{n-1}x \oplus C_n a \oplus C_n b$, with the first, second, and third summands, respectively, belonging to x, a, b. This, larger, complex has a boundary map that acts on a triple by $(c_x, c_a, c_b) \mapsto (-\partial c_x, \partial c_a - c_x, \partial c_b + c_x)$. A quick computation verifies that this is, indeed, a boundary map (i.e. it squares to 0).

Now, in analogy, we would expect a *homotopy* from f_* to g_* to be a chain map $h_*\colon C_* \otimes I_* \to D_*$ such that the component corresponding to a agrees with f_* and the component corresponding with b agrees with g_*. What remains to be settled is where the component corresponding to x should be mapped.

Naming the restrictions $s_n = h_n|_{C_{n-1}x}$, we can reformulate the requirement for being a chain map into a requirement on the family s_n alone. Specifically, the chain map requirement specializes to the requirement that f_* and g_* are chain maps when we are considering the components $C_n a$ and $C_n b$; whereas for the component $C_{n-1}x$ the chain map requirement gives us the equation $-s_*\partial_* v + g_* v - f_* v = \partial_* s_* v$, which we can rewrite as the equation $\partial_* s_* + s_* \partial_* = g_* - f_*$. Hence, we have the following

Definition 3.133   A *chain homotopy* from a chain map $f_* : C_* \to D_*$ to a chain map $g_* : C_* \to D_*$ is a *graded map* $s_* : C_* \to D_*$ of degree 1 such that $\partial_* s_* + s_* \partial_* = g_* - f_*$.

A frequent use for chain homotopies is to show that two maps induce the same map in homology.

Proposition 3.134   *If $f_* : C_* \to D_*$ and $g_* : C_* \to D_*$ are chain homotopic maps then $H(f_*) = H(g_*)$.*

Proof   It is enough to verify that, for any cycle $z \in C_n$, the homology classes $[f_n(z)]$ and $[g_n(z)]$ are equal. The defining equation for a chain homotopy $s_* : f_* \Rightarrow g_*$ reads here

$$\partial_{n+1}(s_n v) + s_{n-1}(\partial_n v) = g_n(v) - f_n(v).$$

This tells us that

$$[g_n(v)] - [f_n(v)] = [g_n(v) - f_n(v)] = [\partial_{n+1}(s_n v) + s_{n-1}(d_n v)] = [\partial_{n+1}(s_n v)] = 0,$$

since $\partial_{n+1}(s_n v) \in B_n$ and v is a cycle.   □

3.3.8   Singular Homology

We have been working with the homology construction on a simplicial complex. Of course, simplicial complexes produce topological spaces. In fact, a topological space can often be given as the geometric realization of many different simplicial complexes. What this means is that if we want to work with a topological space X, we must first construct at least one simplicial complex whose geometric realization is X and, second, convince ourselves that we obtain the same answer, in an appropriate sense, no matter which simplicial complex with geometric realization X we choose. This produces a great deal of "overhead" in the computations, and the second condition is often difficult to verify. A simple and elegant solution to this problem was given by S. Eilenberg (1944), who managed to produce a definition of vector spaces $H_n^{\text{sing}}(X)$ for a space X, without needing to involve a simplicial complex whose geometric realization is homeomorphic to X. Roughly speaking, Eilenberg produced an object much like a simplicial complex with uncountably many vertices and simplices, and defined $H_n^{\text{sing}}(X)$ to be its homology. The construction is known as a *singular homology*. It has the same functoriality properties as those described for a simplicial homology in Proposition 3.132 above but now is defined for topological spaces and maps of topological spaces rather than simplicial complexes. When a space is given as a simplicial complex, the simplicially computed homology agrees with the singular homology in the sense that there is an isomorphism of vector spaces from the simplicial computation to the singular computation, and which respects the linear transformations induced by maps of simplicial complexes. Although there are a number of complicated technical details involved in the construction of a singular homology, one can give a rough summary of the construction as the definition of a simplicial complex with infinitely (in fact uncountably infinitely) many simplices. Roughly speaking, the k-simplices in this complex evaluated on a space X are in bijective correspondence with the continuous maps from the standard k-simplex Δ^k to X. We do not give any further development of singular homology, but we will occasionally allow ourselves to use it. See Hatcher (2002) for the definitions and properties of singular homologies.

3.3.9   Functoriality

We have seen a number of instances of functoriality up to this point. In the last subsection we saw that our intuitive notion of independent cycles can be interpreted as a linear algebraic quantity, namely as the difference in dimensions of certain vector spaces obtained from matrices defined using combinatorial information from the simplicial complex. It turns out, though, that it is useful to consider the dimensions involved and particularly useful to keep track of the vector spaces involved. This is a profound observation, and it was first made in the 1930s by Emmy Noether (Hilton 1988). As we have seen, the formulation of this theory accepted now is that associated with every topological space (or simplicial complex) X and every non-negative integer k is a vector space $H_k(X)$ whose dimension is intuitively interpreted as the number of independent k-dimensional cycles in X, and is therefore called the k-dimensional Betti number of X.

The critical aspect of this definition is that in addition to providing a number which describes an interesting property of the space, it also assigns to every continuous map $f: X \to Y$ of topological spaces a linear transformation between the corresponding homology vector spaces. If one has chosen bases for the relevant vector spaces, one thereby obtains a matrix which is characteristic of f. For example, if we consider the unit circle S^1 in the complex plane, $H_1(S^1)$ is a vector space of dimension 1. If we let f denote the identity map from S^1 to itself, it gives us the 1×1 matrix with entry 1. On the other hand, if we consider the map (again from S^1 to itself) given by complex conjugation, we obtain the 1×1 matrix with entry -1. Thus, in addition to giving us information about spaces themselves, homology also gives us information about continuous maps between spaces.

There is another important notion related to homology, that of *homotopy invariance*. Recall that if $f, g: X \to Y$ are continuous maps of topological spaces then a *homotopy* from f to g is a continuous map $H: X \times [0, 1] \to Y$ such that $H(x, 0) = f(x)$ and $H(x, 1) = g(x)$. Intuitively, this means that there is a continuous family of maps from X to Y, beginning at f and ending at g; the family is $\{H_t\}$, where $H_t(x) = H(x, t)$.

Example 3.135   Suppose that $Y = \mathbb{R}^n$. Then any two maps $f, g: X \to \mathbb{R}^n$ are homotopic, with the homotopy given by $H(x, t) = (1 - t)f(x) + tg(x)$.

Example 3.136   For $X = Y = S^1$, let $f_n: S^1 \to S^1$ be given by $f_n(z) = z^n$, where we think of S^1 as consisting of unit-absolute-value complex numbers. Then f_n is homotopic to f_m if and only if $m = n$.

The *homotopy class* of a continuous map f from X to Y is the family of all maps which are homotopic to f. The set of all continuous maps from X to Y is partitioned into all the different homotopy classes. The classification up to homotopy – by which two different entities are sorted into the same class if they are homotopic – is a discrete classification of the set of all maps. It is quite coarse in general, but it does contain much interesting information. One valuable aspect of homology is that one can study this classification of the continuous maps using it, because of the following property of homology.

Homotopy invariance of homology

If $f, g \colon X \to Y$ are homotopic to each other then the induced maps $H_i(f)$ and $H_i(g)$ are equal for all i. It follows that homotopy equivalent spaces have isomorphic homology, and therefore identical Betti numbers.

This property is quite useful. For example, one can use it to show that certain maps are not homotopic.

Example 3.137   Consider the maps f_n from Example 3.136 above. Then the induced map $H_1(f_n)$ corresponds to multiplication by the integer n. So, if $n \neq m$ then f_n and f_m are not homotopic.

3.3.10   Indirect Methods of Computation

It is impossible to compute directly with the singular model, since it turns out to require working with vector spaces of uncountably infinite dimension. On the other hand, there are numerous more indirect methods that can be used for computation, such as the *Mayer–Vietoris sequence* and the *long exact sequence of a pair*, which often permit the computation of the singular homology of a space X without finding a particular simplicial complex model that is homeomorphic to X. Even in situations where one does have a simplicial complex model for the space, it is often so large that it is necessary to use indirect methods either to complete the calculation or to break it up into more manageable pieces. We give a brief introduction to these methods.

The key notion is that of an *exact sequence of vector spaces*.

Definition 3.138   Let

$$U \xrightarrow{L} V \xrightarrow{M} W$$

be a diagram of linear transformations over a field $\Bbbk$. We say that this sequence of linear transformations is *exact* if (a) $M \circ L \equiv 0$ and (b) the null space of M (which contains the image of L because of requirement (a)) is equal to the image of L. A sequence of linear transformations

$$\cdots \xrightarrow{L_{i+2}} V_{i+1} \xrightarrow{L_{i+1}} V_i \xrightarrow{L_i} V_{i-1} \xrightarrow{L_{i-1}} \cdots$$

of arbitrary length or even of infinite length is said to be exact if and only if each three-term sequence

$$V_{i+1} \xrightarrow{L_{i+1}} V_i \xrightarrow{L_i} V_{i-1}$$

is exact.

The following is a numerical consequence of exactness.

Proposition 3.139   *A sequence of linear transformations*

$$V_{i+1} \xrightarrow{L_{i+1}} V_i \xrightarrow{L_i} V_{i-1}$$

is exact if and only if $\dim(V_i) = \text{rank}(L_{i+1}) + \text{rank}(L_i)$ *for all i. Similarly, if a sequence of linear transformations*

$$V_{i+2} \xrightarrow{L} V_{i+1} \longrightarrow V_i \longrightarrow V_{i-1} \xrightarrow{L'} V_{i-2}$$

is exact then we have the equation

$$\dim V_i = \dim V_{i-1} + \dim V_{i+1} - \text{rank } L - \text{rank } L'.$$

This result is often used, when one has knowledge of $\dim(V_{i+1})$ and $\dim(V_{i-1})$ together with the ranks of L_{i+2} and L_{i-1}, to evaluate $\dim(V_i)$, as we shall see.

We now describe the *Mayer–Vietoris* long exact sequence. It arises when we have a simplicial complex X which is the union of two subcomplexes Y and Z. In this case, $Y \cap Z$ is also a subcomplex of X. The goal of the Mayer–Vietoris sequence is to obtain a long exact sequence in which the homology groups of Y, Z, and $Y \cap Z$ are involved in addition to the homology group of X, and in which one can therefore often infer the homology of X from that of Y, Z, and $Y \cap Z$.

Theorem 3.140 *With X, Y, and Z as above, there is a long exact sequence*

$$\cdots \longrightarrow H_i(Y \cap Z) \xrightarrow{\alpha_i} H_i(Y) \oplus H_i(Z) \longrightarrow H_i(X) \xrightarrow{\delta}$$

$$\xrightarrow{\delta} H_{i-1}(Y \cap Z) \xrightarrow{\alpha_{i-1}} H_{i-1}(Y) \oplus H_{i-1}(Z) \longrightarrow \cdots.$$

The sequence continues, with period 3 and a dimension shift, indefinitely for all non-negative values of i. It terminates on the right at $H_0(X)$; the groups to the right of that are identically zero. The linear transformation α_i is given by $\alpha_i(\xi) = (H_i(i_0)(\xi), -H_i(i_1)(\xi)$, where $i_0 : Y \cap Z \hookrightarrow Y$ and $i_1 : Y \cap Z \hookrightarrow Z$ are the inclusions. The transformation δ is somewhat complicated to define and we refer to Hatcher (2002). This sequence also exists for the singular homology when Y and Z form an open covering of a space X.

This result asserts that if we have complete understanding of the homology groups of Y, Z, and $Y \cap Z$, together with the linear transformations $H_k(i_0)$ and $H_k(i_1)$, then we can evaluate the dimensions of the homology groups of X. This kind of computation is often referred to as a *local to global* result. The precise statement in terms of dimensions of homology groups and ranks of linear transformations is the following.

Corollary 3.141 *Let $\alpha_i : H_i(Y \cap Z) \longrightarrow H_i(Y) \oplus H_i(Z)$ denote the linear transformation $\alpha_i(\xi) = (H_i(i_0)(\xi), -H_i(i_1)(\xi))$. Then the dimension of $H_i(X)$ is*

$$\dim(H_i(Y)) + \dim(H_i(Z)) + \dim(H_{i-1}(Y \cap Z)) - \text{rank}(\alpha_i) - \text{rank}(\alpha_{i-1}).$$

Example 3.142 We can show how this result allows us to compute the singular homology of the circle S^1 without finding a simplicial complex model for it. Suppose that we cover S^1 by the two subsets $U = S^1 - \{(0, -1)\}$ and $V = S^1 - \{(0, 1)\}$. It is clear that U and V are homeomorphic to the open intervals $(-\frac{\pi}{2}, \frac{3\pi}{2})$ and $(\frac{\pi}{2}, \frac{5\pi}{2})$, respectively, using the polar coordinate parametrization of the circle. They are therefore both contractible, i.e. homotopy equivalent to a single point. Consequently, $H_i(U) = 0$ and $H_i(V) = 0$ for $i > 0$, and $H_0(U) \cong H_0(V) \cong \Bbbk$. On the other hand, $U \cap V = S^1 - \{\pm 1\}$ is homeomorphic to a disjoint union of two intervals, namely $(-\frac{\pi}{2}, \frac{\pi}{2})$ and $(\frac{\pi}{2}, \frac{3\pi}{2})$, and is therefore homotopy equivalent to the union of two discrete points. It follows that $H_i(U \cap V) \cong 0$ for $i > 0$ and that $H_0(U \cap V) \cong \Bbbk \oplus \Bbbk$. This means that the long exact sequence looks as follows:

$$\longrightarrow 0 \xrightarrow{f} H_1(X) \longrightarrow H_0(U \cap V) \cong \Bbbk \oplus \Bbbk \xrightarrow{g} H_0(U) \oplus H_0(V) \cong \Bbbk \oplus \Bbbk \longrightarrow \Bbbk \longrightarrow 0.$$

The two connected components of $U \cap V$ form a basis for $H_0(U \cap V)$, and the single connected components of U and V form a basis for $H_0(U) \oplus H_0(V)$; in this basis it is readily computed that the matrix relative to the basis takes the form

$$\begin{bmatrix} 1 & 1 \\ -1 & -1 \end{bmatrix}$$

Since this matrix has rank 1, the formula from Corollary 3.141 is as follows:

$$\dim(H_1(X)) = (\dim(H_1(U)) + \dim(H_1(V))) + \dim(H_0(U \cap V)) - \text{rank}(f) - \text{rank}(g)$$

or

$$\dim(H_1(X)) = (0 + 0) + 2 - 0 - 1 = 1.$$

The long exact sequence of a pair is another powerful method for computing the Betti numbers of simplicial complexes and of spaces, using singular homology. It applies to the situation where we have a simplicial complex X and a subcomplex Y. We first discuss *relative homology*. We have the chain complexes $C_*(X)$ and $C_*(Y)$, with an inclusion (which is a chain map) $C_*(Y) \hookrightarrow C_*(X)$, and we may form quotient vector spaces $C_i(X)/C_i(Y)$, which we will denote by $C_i(X, Y)$. It follows from standard algebra that the boundary maps in $C_*(X)$ induce boundary maps $\partial: C_i(X, Y) \longrightarrow C_{i-1}(X, Y)$. With these boundary maps we may evaluate its homology, which we denote by $H_i(X, Y)$ and which we refer to as the *relative homology of X with respect to Y*. There is now the following exact sequence of vector spaces, which is a powerful tool for computing singular homology as well as simplicial homology.

Proposition 3.143   *Given X and Y as above, we have a long exact sequence*

$$\cdots \longrightarrow H_{i+1}(X, Y) \longrightarrow H_i(Y) \longrightarrow H_i(X) \longrightarrow H_i(X, Y) \longrightarrow H_{i-1}(Y) \longrightarrow \cdots.$$

There is an identical sequence for a singular homology where one is also able to define $H_i(X, Y)$ for any space X and subspace $Y \subseteq X$.

The power of this method comes from the *excision property* for relative homology. Consider again a simplicial complex X, a subcomplex $Y \subseteq X$, and a second subcomplex $Z \subseteq Y$. We may form the subcomplex $Y - Z$ by declaring that a simplex of Y is in $Y - Z$ if and only if it is a face of a simplex which is not contained in Z. We can also construct $X - Z$ in the same way, and of course $Y - Z$ is a subcomplex of $X - Z$.

Proposition 3.144   *Let $Z \subseteq Y \subseteq X$ be inclusions of subcomplexes. Then the natural chain map $C_*(X - Z, Y - Z) \hookrightarrow C_*(X, Y)$ induces an isomorphism*

$$H_i(X - Z, Y - Z) \xrightarrow{\sim} H_i(X, Y).$$

There is an analogous statement for a singular homology, where we have subspaces
$Z \subseteq Y \subseteq X$ *and where Z is a subset contained in a closed subset of X contained in Y*
and Y is open.

We now give an example of how this method is used.

Example 3.145   We will compute inductively the singular homology of the n-sphere
S^n using this method. We begin with the case of S^0, which is the union of two discrete
points, and therefore its homology is given by $H_0(S^0) \cong \Bbbk \oplus \Bbbk$ and $H_i(S^0) \cong 0$. Next,
we consider the circle $S^1 = X$, and let $Y = X - \{(1, 0)\}$ and $Z = \{(x, y) \in X \mid x \leq 0\}$.

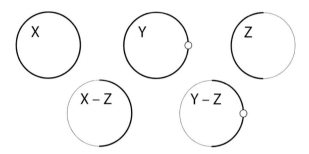

We do not have a direct understanding of $H_*(X, Y)$, but using the excision result in
Proposition 3.144 we conclude that $H_*(X, Y) \cong H_*(X - Z, Y - Z)$, and we can obtain
information about $H_*(X - Z, Y - Z)$ using the long exact sequence of a pair, as follows.
The space $X - Z$ is obtained by removing from the circle all the points with non-positive
x-coordinate. This space is an open arc and is therefore homeomorphic to an open
interval. An open interval is homotopy equivalent to a point, and therefore we have that
$H_0(X - Z) \cong \Bbbk$ and $H_i(X - Z) \cong 0$ for $i > 0$. On the other hand, the space $Y - Z$ is
homeomorphic to a disjoint union of two intervals, which is homotopy equivalent to two
discrete points, i.e. to S^0. We conclude that $H_0(Y - Z) \cong \Bbbk \oplus \Bbbk$ and that $H_i(Y - Z) \cong 0$
for $i > 0$. We have the long exact sequence of the pair $(X - Z, Y - Z)$, which looks as
follows in its low degrees:

$$H_1(X - Z) \longrightarrow H_1(X - Z, Y - Z) \longrightarrow H_0(Y - Z) \longrightarrow H_0(X - Z)$$
$$\longrightarrow H_0(X - Z, Y - Z) \longrightarrow 0.$$

Since $H_1(X - Z) \cong 0$, it follows from the second numerical criterion in Proposition
3.139 that the dimension of $H_1(X - Z, Y - Z)$ is equal to the difference

$$\dim H_0(Y - Z) - \text{rank}(H_0(Y - Z) \to H_0(X - Z)) = 2 - 1 = 1.$$

Therefore, the excision theorem tells us that $H_1(X, Y)$ has dimension 1. It is easy to
verify that all $H_i(X, Y) \cong H_i(X - Z, Y - Z)$ vanish for $i \neq 1$, since, for example, in
higher degrees $H_i(X - Z, Y - Z)$ is trapped between vanishing groups. We now consider
the long exact sequence of the pair (X, Y):

$$\cdots \longrightarrow H_1(Y) \longrightarrow H_1(X) \longrightarrow H_1(X, Y) \longrightarrow H_1(Y) \longrightarrow H_1(X) \cdots .$$

Given that we know that $H_1(Y) \cong 0$, it follows that the dimension of $H_1(X)$ is equal to the dimension of $H_1(X, Y)$, which is equal to 1, minus the rank of the linear transformation $H_1(X, Y) \longrightarrow H_0(Y)$. We claim this rank is 0, for $H_0(Y) \cong H_0(X) \cong \Bbbk$, and the transformation $H_0(Y) \longrightarrow H_0(X)$ is an isomorphism sending the basis element corresponding to the unique connected component of Y to the unique connected component X. The exactness of the sequence

$$H_1(X, Y) \longrightarrow H_0(Y) \longrightarrow H_0(X)$$

guarantees that the transformation $H_1(X, Y) \longrightarrow H_0(Y)$ vanishes and therefore has rank 0. It follows that $H_1(X)$ has dimension 1. One can now proceed inductively to prove that the dimension of $H_i(S^n)$, $\dim H_i(S^n)$, equals 1 for $i = 0, n$ and equals 0 otherwise. Let X denote the unit sphere in $\mathbb{R}^{n+1}$, Y the set $S^n - \{(1, \ldots, 0)\}$, and Z the set $\{\vec{x} \in S^n \mid x_0 \leq 0\}$, where $\mathbb{R}^n$ is given coordinates $x_0, x_1, \ldots, x_n$. We can now proceed inductively, since it turns out that $Y - Z$ is homotopy equivalent to S^{n-1} and $X - Z$ is contractible, from which we are able to conclude that $H_n(X - Z, Y - Z)$ has dimension $= 1$. The excision theorem then guarantees that $H_n(X, Y)$ also has dimension 1. It also turns out that Y is contractible, from which it follows that $H_n(S^n) \cong H_n(X, Y)$, and therefore has dimension 1. The cases $i \neq n$ are easy to check as well, and one obtains a calculation of $H_i(S^n)$ for all i.

3.3.11   Importance of Functoriality

Functoriality is defined as the study of maps of mathematical objects as opposed to the study of mathematical objects in isolation. It is perhaps the most characteristic feature of the methods developed in algebraic topology. We introduce some examples which illustrate its significance.

1. In Section 3.3.10 above we introduced methods for computing homologies which apply even for singular homologies, and which permit parallelization of the computation of simplicial homologies. These methods depend crucially on the functoriality of the homology construction, since one ingredient in the formulas derived from them is the rank of induced maps on the homology. There are many generalizations of the methods of Section 3.3.10, and virtually all of them depend in a similar way on the analysis of induced linear transformations.

2. Functoriality is also critical in most applications of topology within mathematics or outside it. For example, the *Lefschetz fixed point theorem* provides a way of guaranteeing the existence of a fixed point of a self-map $f : X \to X$, where X is a space which is homeomorphic to a simplicial complex. The statement of this theorem is as follows.

   Let X be a space homeomorphic to a finite simplicial complex. For each $k \geq 0$, we have an induced linear transformation $H_k(f, \mathbb{Q})$, where $\mathbb{Q}$ denotes the field of rational numbers. The *Lefschetz number* of f is defined to be the alternating sum

   $$\Lambda_f = \sum_i (-1)^i \operatorname{Tr} H_i(f, \mathbb{Q}),$$

where Tr denotes the trace of the linear transformation. Because the trace depends only on the conjugacy class of a matrix, it is an invariant of the linear transformation: it is independent of the choice of basis and therefore the matrix representing it.

Theorem 3.146 *For $f: X \to X$ as above, suppose that $\Lambda_f \neq 0$. Then f has at least one fixed point.*

3. In statistics, it is frequently useful to define methods of automatically clustering a data set, i.e. of finding a way of dividing it up into conceptually coherent pieces. For example, if we are given a data set with a notion of distance (i.e. a finite metric space), one can perform so-called *single-linkage clustering* with threshold parameter ϵ by creating a graph (a one-dimensional simplicial complex) whose nodes are the points in the metric space and where an edge is included exactly if the two points defining it are a distance $\leq \epsilon$ apart and by then partitioning the metric space into the components of the associated graph. One problem with this procedure is that it requires a choice of ϵ, which is quite problematic. A solution to this difficulty is the notion of *hierarchical clustering*, which provides as output not a single partition but a nested family of partitions. The key point is that there an induced map from the set of clusters at threshold value ϵ to the set of clusters at threshold value ϵ' when $\epsilon \leq \epsilon'$, and we can view this as a functoriality statement for the set of graph components rather than for homology. This nested family can then be conveniently visualized via a dendrogram, i.e a tree with a reference map to the non-negative real line. This construction provides a summary of the behavior of all the clusterings at all values of the scale parameter ϵ, and is a useful guide to choosing values which are likely to produce interesting results. Of course, the dendrogram contains information about clusterings (i.e. sets of clusters), but it also demonstrates the relationships between clusterings at different choices of ϵ. Persistent homology, which we will introduce in the next chapter, also depends on a threshold parameter and relies for its definition on a similar functoriality statement about homologies at two different threshold values.

4. Often spaces are equipped with a group of symmetries. When this is the case, there is a notion of an *equivariant homology* which contains information not only about the space but also about the action of symmetry groups such as fixed point sets of various elements. The definition of an equivariant homology is entirely dependent on functoriality statements for the elements of the symmetry group. This is a very rich area of study, with numerous applications within mathematical physics. See Szabo (2000) for some examples.

5. It is also interesting to study the topology of spaces X equipped with a reference map to a base space B. This area of study is called *parametrized topology*, and clearly also depends in a crucial way on the induced map on the homology for the reference map. A striking application of these methods in number theory and algebraic geometry can be found in Artin & Mazur (1986) and Freitag & Kiehl (1988).

4   Shape of Data

For the methods we develop here, the most important aspect of a data set is that it quantifies in some way the difference between observations; thus we will regard a *data set* as a finite collection of observations characterized by some dissimilarity measure. Preferably this measure would be a metric, but this requirement can be relaxed for some of our methods.

We will start out by describing this perspective, which we think of as the point cloud view of data, and then continue with a range of useful constructions that let us leverage homology and functoriality to produce topological methods of data analysis.

4.1   Zero-Dimensional Topology: Single-Linkage Clustering

The simplest aspect of the shape of a geometric object is its number of connected components. Statisticians have thought a great deal about what the counterpart to connected components should be for point cloud data, under the heading of *clustering* (see Hartigan 1975; Kogan 2007). One scheme for clustering, called *single-linkage hierarchical clustering*, proceeds as follows. We suppose that we are given a finite dissimilarity space with points $X = \{x_1, \ldots, x_n\}$ and pairwise dissimilarities $\mathfrak{D}(x_i, x_j)$. For every non-negative threshhold R, we may form a relation $\sim_R$ on the set X by the criterion

$$x \sim_R x' \quad \text{if and only if} \quad \mathfrak{D}(x, x') \leq R.$$

We let $\simeq_R$ denote the equivalence relation generated by $\sim_R$. The set of equivalence classes under $\simeq_R$ now gives a partition of X, which can be thought of as a candidate for the connected components in X. So, for each threshold R, we obtain a partition of X. One can now ask which choice of R is the "right" one. This is an ill-defined question, although there are interesting heuristics. Another approach is to observe that there is compatibility across changes in R, in that if $R \leq R'$ then the partition associated with R' is coarser than the partition associated with R. This means that clustered points in R stay clustered in R', as is indicated in Figure 4.1. The diagram indicates the change in clustering as the threshold is altered, and shows the increasing coarseness as R increases. What has been recognized by statisticians is that there is a single profile, called a *dendrogram*, which encodes the clusterings at all the thresholds simultaneously. Figure 4.2 illustrates the dendrogram associated with the situation given above. The result is a tree, T

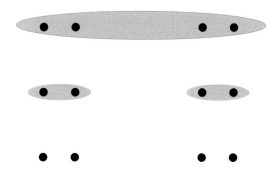

Figure 4.1 Single-linkage hierarchical clustering, in which the distance parameter R increases upwards, giving coarser clusters

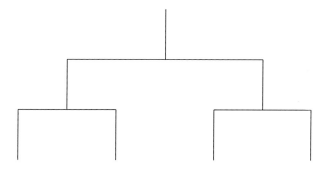

Figure 4.2 Dendrogram

(a one-dimensional simplicial complex with no loops) together with a reference map from T to the non-negative real line. This tree can be viewed all at once. The clustering at a given threshold R is given by drawing a horizontal line at level R across the tree, and the clusters correspond to the points of intersection. In the above picture, the reference map is the height function above the x-axis.

There is another way to interpret the dendrogram that is less visual but formally identical. For each threshhold R, we let X_R denote the set of equivalence classes for the equivalence relation $\approx_R$. Because the partition only coarsens as R increases, there is a map of sets $X_R \to X_{R'}$ whenever $R \le R'$, which assigns to each cluster at level R the (unique) cluster at level R' in which it is included. This construction is sufficiently useful that we will give it a name.

Definition 4.1 By a *persistent set*, we mean a family of sets $\{X_R\}_{R \in \mathbb{R}}$ together with set maps $\varphi_R^{R'}: X_R \to X_{R'}$ for all $R \le R'$, such that $\varphi_{R'}^{R''} \varphi_R^{R'} = \varphi_R^{R''}$ for all $R \le R' \le R''$. More generally, for any kinds of objects, such as simplicial complexes, vector spaces, topological spaces, etc., we may speak of a persistent object as a family of such objects parametrized by $\mathbb{R}$, together with maps (maps of simplicial complexes, linear transformations, continuous maps, etc.) from an object parametrized by r to one parametrized by r', whenever $r \le r'$, with the same compatibilities as those mentioned above.

For each persistent set, there is an associated dendrogram and vice versa. There is a reformulation using topological notions of the dendrogram (and hence of the persistent set) associated with a finite metric space, which will make much clearer the development of methods of defining higher-dimensional homology for finite metric spaces.

———————————————— For the advanced reader ————————————————

The advanced reader may well recognize the structure of Definition 4.1. The requirement on composing set maps means that a persistent set is a functor $(\mathbb{R}, \leq) \rightarrow \text{Set}$, or equivalently a set-valued pre-sheaf over $(\mathbb{R}, \leq)$. In this way, persistent objects in other categories can be viewed simply as pre-sheaves valued in those categories, which is a perspective that is producing much interesting research and many insights.

It turns out that hierarchical single-linkage clustering, although its construction is very natural, often produces results that exhibit a *chaining phenomenon*, which is for example exhibited by the merging of a long list of distinct and very small clusters into a single large cluster. It turns out that one can define other kinds of linkage clustering models as follows.

Definition 4.2   By a *linkage function*, we mean any positive function $\mathfrak{L}$ on the set of isomorphism classes of objects of the form (D, D_0, D_1), where D is a finite dissimilarity space and D_0 and D_1 are disjoint subsets of D such that $D = D_0 \cup D_1$. An isomorphism from (D, D_0, D_1) to (D', D_0', D_1') is an isomorphism of the dissimilarity space D to D' that carries D_i to D_i' for $i = 0, 1$.

Here are three commonly occurring linkage functions $\mathfrak{L}$.

1. $\mathfrak{L}^{\min}(D, D_0, D_1) = $ minimum distance from points of D_0 to points of D_1.
2. $\mathfrak{L}^{\max}(D, D_0, D_1) = $ maximum distance from points of D_0 to points of D_1.
3. $\mathfrak{L}^{\text{ave}}(D, D_0, D_1) = $ average distance from points of D_0 to points of D_1.

A linkage function can be used to create generalized versions of hierarchical single-linkage clustering.

Definition 4.3   By a *hierarchical clustering* of a finite dissimilarity space $(X, \mathfrak{D})$, we mean a family of equivalence relations $\{E_r\}_r$ of the set X parametrized by a real-valued parameter r satisfying the following properties.

1. For $r \leq 0$, the equivalence relation E_r is the discrete equivalence relation $\Delta \subseteq X \times X$, where $\Delta = \{(x, x) | x \in X\}$.
2. For $r \leq r'$, we have that $E_r \subseteq E_{r'}$.
3. Given a sequence of real numbers $r_0 \geq r_1 \geq r_2 \geq \cdots$ that is bounded below and has the property that all the equivalence relations E_{r_i} are equal to a fixed equivalence relation R, then the equivalence relation E_ρ is equal to E, where $\rho = \inf_i r_i$.

Any hierarchical clustering of $(X, \mathfrak{D})$ produces a dendrogram, just as in the case of hierarchical single-linkage clustering considered above. We first observe that the finiteness of X means that any for any hierarchical clustering of X, we have a non-negative integer k, a finite sequence of real numbers $r_0 < r_1 < \cdots < r_k$, and equivalence relations E_i satisfying the following properties.

1. $r_0 = 0$.
2. For all $r \in [r_i, r_{i+1})$, $E_r = E_{r_i}$.
3. E_{r_i} is strictly finer than $E_{r_{i+1}}$ for $i < k$.
4. $E_r = E_{r_k}$ for all $r \geq r_k$.

Suppose that we are given a finite set X, an equivalence relation E on X, and a relation R on the set of equivalence classes of E. Then by the *expansion of E along R* we mean the equivalence relation $E[R]$ on X defined by

$$(x, x') \in E[R] \quad \text{if and only if} \quad ([x], [x']) \in \tau(R).$$

The relation E is weakly finer than $E[R]$: that is, E is equal to or finer than $E[R]$.

Suppose we are given a linkage function $\mathfrak{L}$ and a dissimilarity space $(X, \mathfrak{D})$. We produce a sequence of equivalence relations on X using the following algorithm.

1. Set i = 0.
2. Set $\mathfrak{E}_i$ equal to the discrete equivalence relation on X.
3. Output $\mathfrak{E}_i$.
4. Set $\mathfrak{D}_i$ equal to the set of equivalence classes of $\mathfrak{E}_i$.
5. Is $\#(\mathfrak{D}_i) = 1$? If yes, set $k = i$ and quit. If no, proceed.
6. Set S_i equal to the set of pairs (ξ, ξ'), where ξ and ξ' are equivalence classes under $\mathfrak{E}_i$ and $\xi \neq \xi'$.
7. Set $\lambda_i = \min_{(x, x') \in S_i} \mathfrak{L}(\xi, \xi')$. Output λ_i.
8. Set R_i equal to the relation $\{(\xi, \xi') \in S_i | \mathfrak{L}(\xi, \xi') \leq \lambda_i\}$.
9. Set $\mathfrak{E}_{i+1} = \mathfrak{E}_i[R_i]$.
10. Set i = i + 1.
11. Go to step 4.

The output consists of a sequence of equivalence relations $\{\mathfrak{E}_0, \mathfrak{E}_1, \ldots, \mathfrak{E}_k\}$ and a sequence of numbers $\lambda_0 < \lambda_1 < \cdots < \lambda_{k-1}$. From this data we can construct a hierarchical clustering $\{E_r\}$ of X by the following requirements.

1. E_r is the discrete equivalence relation if $r < \lambda_0$.
2. $E_r = \mathfrak{E}_i$ for $\lambda_{i-1} \leq r < \lambda_i$.
3. E_r is the indiscrete equivalence relation for $r \geq \lambda_{k-1}$.

The clustering algorithms corresponding to the linkage functions $\mathfrak{L}^{\min}$, $\mathfrak{L}^{\max}$, and $\mathfrak{L}^{\text{ave}}$ are called single-linkage, complete-linkage, and average-linkage hierarchical clustering algorithms, respectively.

4.2   The Nerve Construction and Soft Clustering

We looked at clustering schemes in Sections 2.2 and 4.1. Clustering schemes begin with point cloud data and from it produce a partition of the data, i.e. a decomposition of the set of points into a union of disjoint subsets. Often, though, one finds that the clustering algorithm appears to be forcing some very close points into different clusters. Figure 4.3 shows one such example, derived from Faúndez-Abans et al. (1996). The points in question are nebulae, and the different types of nebulae are denoted by the shapes of the dots, i.e. squares, triangles, stars, and crosses. They are plotted on a table with two

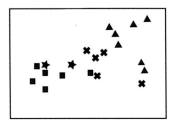

Figure 4.3 Cluster analysis for nebulae

features of nebulae as the two coordinate axes. It is clear that the different groups do not separate very strongly, and that one might better consider them in terms of a *soft clustering*, where the various subsets are permitted to overlap. For any set X and finite collection of non-empty subsets $\mathcal{U}$ of X, such that

$$X = \bigcup_{U \in \mathcal{U}} U,$$

we refer to the collection $\mathcal{U}$ as a *covering* of X. The question is how to encode the information about the possible intersection of elements of $\mathcal{U}$ in a simple structure. Recall from Section 3.3 the notion of an abstract simplicial complex, and suppose that the covering $\mathcal{U}$ is given by

$$X = \bigcup_{i=1}^{n} U_i.$$

Definition 4.4   By the *nerve* of $\mathcal{U}$, we mean the abstract simplicial complex $(V_{\mathcal{U}}, \Sigma_{\mathcal{U}})$, where $V_{\mathcal{U}} = \{1, \ldots n\}$ and $\Sigma_{\mathcal{U}}$ consists of all non-empty collections $\{i_0, \ldots, i_s\} \subseteq \{1, \ldots, n\}$ such that

$$U_{i_0} \cap \cdots \cap U_{i_s} \neq \emptyset.$$

We denote the nerve construction $(V_{\mathcal{U}}, \Sigma_{\mathcal{U}})$ by $N(\mathcal{U})$.

Suppose the covering $\mathcal{U}$ is a partition, so that it corresponds to a clustering. In this case, the intersections of distinct subsets of $\mathcal{U}$ are all empty and $N(\mathcal{U})$ is a zero-dimensional complex, i.e. it is the discrete set whose elements are the distinct elements of $\mathcal{U}$, and there are no edges connecting any of distinct pairs of them. In the context of data, this will mean that the vertices are in one-to-one correspondence with the clusters.

Example 4.5 In Figure 4.4 below, on the left, we see a covering of a set that is a "fattened" version of the boundary of a square. It is covered by four sets, two of which are colored in red and two of which are colored in blue. They intersect in four purple sets which lie at the corners.

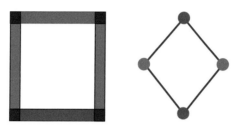

Figure 4.4 A covering of a set (left) and the corresponding nerve (right) of a covering

The nerve is obtained by assigning a node to each set in the covering, and connecting nodes by edges corresponding to the intersections among the sets in the covering. One can verify that the original set (regarded as a space) is homotopy equivalent to the geometric realization of the nerve complex. This is not an accident, as we shall see in the following lemma.

Suppose that X is a topological space, and that we are given a covering $\mathcal{U} = \{U_1, \ldots, U_n\}$ such that U_i is an open set in X. Then $\mathcal{U}$ is said to be an *open covering* of X.

Lemma 4.6 (Nerve lemma) *Suppose we are given an open covering* $\mathcal{U} = \{U_1, \ldots, U_n\}$ *of a topological space* X *that is a subset of* $\mathbb{R}^n$. *Suppose further that each of the intersections*

$$U_{i_1} \cap \cdots \cap U_{i_s}$$

is either empty or contractible (i.e. homotopy equivalent to the one-point space). Then X *is homotopy equivalent to* $N(\mathcal{U})$.

Remark 4.7 The requirement that X be a subset of $\mathbb{R}^n$ can be relaxed to *paracompact spaces*; see Hatcher (2002), where it is stated as Corollary 4G.3. In particular, it holds for all metric spaces. The nerve lemma was originally proved in Borsuk (1948).

Here are some examples of this construction.

Example 4.8 Consider the circle, with covering $\mathcal{U} = \{U_0, U_1, U_2\}$ defined by

$$U_0 = \left\{ \theta \,\middle|\, 0 \leq \theta \leq \frac{2\pi}{3} \right\},$$

$$U_1 = \left\{ \theta \,\middle|\, \frac{2\pi}{3} \leq \theta \leq \frac{4\pi}{3} \right\},$$

$$U_2 = \left\{ \theta \,\middle|\, \frac{4\pi}{3} \leq \theta \leq 2\pi \right\}.$$

Note that $U_0 \cap U_1 \neq \emptyset$, $U_1 \cap U_2 \neq \emptyset$, and $U_0 \cap U_2 \neq \emptyset$, but that $U_0 \cap U_1 \cap U_2 = \emptyset$. Consequently, the nerve of $\mathcal{U}$ has three vertices, corresponding to U_0, U_1, and U_2, with an edge between any pair of vertices, but no 2-simplex. The nerve is in this case the boundary of a triangle.

Example 4.9   Let X denote the boundary of the unit cube in $\mathbb{R}^3$, with a covering $\mathcal{U}$ given by the six faces of the cube. Let $U_{0,1}$ denote the faces given by $x = 0$ or 1; similarly, $V_{0,1}$ for y and $W_{0,1}$ for z. We now consider the intersections. We note that no family of elements which that contains two sets corresponding to the same letter (U, V, or W) will have a non-empty intersection, since the coordinate corresponding to that letter would be required to have the values 0 and 1 simultaneously. On the other hand, any pairwise intersection which does not have two occurrences of the same letter has non-empty intersection. Consequently, we have 12 such pairwise intersections, given explicitly by the list

$$\{U_0 \cap V_0, \ U_0 \cap V_1, \ U_1 \cap V_0, \ U_1 \cap V_1, \ U_0 \cap W_0, \ U_0 \cap W_1,$$
$$U_1 \cap W_0, \ U_1 \cap W_1, \ V_0 \cap W_0, \ V_0 \cap W_1, \ V_1 \cap W_0, \ V_1 \cap W_1\}.$$

These intersections correspond to the edges of the cube, of which there are 12. We similarly have eight non-trivial threefold intersections, parametrized by the choice of 0 or 1 as the subscript for each of U, V, and W. These correspond to the eight vertices of the cube. The nerve of $\mathcal{U}$ will therefore be a simplicial complex with six vertices, 12, and eight triangles. It is an *octahedron*.

We note that in both examples above, the nerve of the covering is homeomorphic to the underlying space, since a triangle is homeomorphic to a circle and the surface of the octahedron and the surface of the cube are both homeomorphic to the sphere, and hence to each other. This is of course not generally true.

Example 4.10 Consider the covering $\mathcal{U}$ of the circle by two sets shown below.

The two sets are in red and blue, and the intersection is the two black points. The nerve of $\mathcal{U}$ has two vertices corresponding to the two sets, and the two sets intersect, so they span one simplex. The nerve is the interval (the topological space homeomorphic to $[0, 1]$), which is not homeomorphic to a circle. We can tell this, for example, by observing that removing any single point from the circle leaves a connected space, homeomorphic to an interval, while removing an interior point from an interval leaves a disconnected space.

This particular example can be resolved, as in Carrière & Oudot (2018), by introducing the *multi-nerve*: instead of a simplicial complex, create a *semi-simplicial set* where a disconnected intersection produces one separate simplex for each connected component in the intersection. The multi-nerve construction gives an elegant way of handling situations where the intersections of cover elements fail to be contractible through being disconnected, but not situations where the contractibility fails through higher-dimensional structures. One approach to statistically testing a covering for the possibility of the nerve lemma failing was developed in Vejdemo-Johansson & Leshchenko (2020).
We will also be working with relationships between coverings.

Definition 4.11 A *covered set* is a pair $(X, \mathcal{U})$, where X is a set and $\mathcal{U} = \{U_\alpha\}_{\alpha \in A}$ is a finite covering of X. Note that the vertex set of $N\mathcal{U}$ is A. A map $\Theta \colon (X, \mathcal{U}) \to (Y, \mathfrak{V})$ is a pair (θ, η), where $\theta \colon X \to Y$ is a map of sets and $\eta \colon A \to B$, with the property that $f(U_\alpha) \subseteq V_\beta$ for all $\alpha \in A$. Here $\mathcal{U} = \{U_\alpha\}_{\alpha \in A}$ and $\mathfrak{V} = \{V_\beta\}_{\beta \in B}$.

Proposition 4.12 Let $\Theta = (\theta, \eta) \colon (X, \mathcal{U}) \to (Y, \mathfrak{V})$ denote a map of covered sets, where again $\mathcal{U} = \{U_\alpha\}_{\alpha \in A}$ and $\mathfrak{V} = \{V_\beta\}_{\beta \in B}$. Then there is a unique map of simplicial complexes

$$N\Theta \colon N\mathcal{U} \longrightarrow N\mathfrak{V}$$

such that, on vertices, $N\Theta = \eta$.

Most of the constructions of simplicial complexes from point clouds proceed by constructing a covering or a family of coverings and applying the nerve construction to that covering or family of coverings.

4.3   Complexes for Point Cloud Data

The basic approach to defining topological invariants of point cloud data is to assign a persistent simplicial complex to the cloud and *define* the invariants of the cloud to be the invariants of this associated complex. The simplicial complexes associated with the cloud depend on a scale parameter ϵ. In this section, we will generally assume that the scale is fixed and look at various ways of associating a complex with he cloud and the interrelationships between these different approaches. In the following chapter, we will begin to study invariants which are independent of scale formed by combining complexes as the scale varies.

As a consistency check, when a point cloud is obtained by sufficiently dense sampling from a suitable topological space (e.g. a Riemannian manifold) one would like to know whether the invariants of the cloud agree with the invariants of the space with high probability. More generally, one would like to verify that if the samples are taken in an ambient metric space (e.g. Euclidean space) and are "near" to an underlying manifold then the invariants of the manifold and the point cloud coincide with high probability. The central theoretical work in this chapter is aimed at establishing these sorts of consistency results.

4.3.1   The Čech Complex

Suppose we are given a point cloud $Z \subset \mathbb{R}^n$. For a fixed ϵ, we define the covering $\mathfrak{U}_\epsilon^{\mathrm{Cech}}$ to be the family $\{B(z, \epsilon)\}_{z \in Z}$, i.e. we take Euclidean balls of radius ϵ around the points of Z.

Definition 4.13   The Čech complex of the point cloud $Z \subset \mathbb{R}^n$ at scale ϵ is defined to be the nerve $N(\mathfrak{U}_\epsilon^{\mathrm{Cech}})$, and we will denote the complex by $C^{\mathrm{Cech}}(Z, \epsilon)$. The embedding $Z \hookrightarrow \mathbb{R}^n$ is understood, and is not included in the notation.

Example 4.14   In Figure 4.5 below, we show the Čech complex of the point cloud $\left\{(\pm 1, 0), (0, \pm 1), \left(\pm \frac{\sqrt{2}}{2}, \pm \frac{\sqrt{2}}{2}\right)\right\}$ at scale $\epsilon = \frac{1}{2}$.

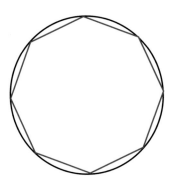

Figure 4.5 Čech complex (red octagon) of a small point cloud

More generally, we can define the Čech complex associated with a point cloud Z contained in any metric space (E, ∂), such as a Riemannian manifold, in an analogous fashion. In this case, it is necessary to include the ambient metric space E in the notation, and we write $C^{\text{Cech}}(Z, E, \epsilon)$ for the Čech construction.

Remark 4.15 An important property of the Čech complex is that is equipped with certain theoretical guarantees in the case of compact Riemannian manifolds (see Bishop & Crittenden 1964 for a discussion of Riemannian manifolds). A subset C of a Riemannian manifold M is said to be *geodesically convex* if any two points in C are connected by a unique geodesic contained entirely in C. It can be shown that any geodesically convex open set is homeomorphic to a ball in $\mathbb{R}^n$, where n is the dimension of M, and is therefore contractible. It is immediate from the definition that finite intersections of geodesically convex sets are also geodesically convex. The following theorem is fundamental, and is proved in Bishop & Crittenden (1964).

Theorem 4.16 *For any compact Riemannian manifold M, there is an $\epsilon > 0$ such that every geodesic ball of radius $\leq \epsilon$ is geodesically convex.*

We say that a finite subset $X \subseteq M$, where M is a compact Riemannian manifold, is an ϵ-net if for every point $m \in M$ there is a point $x \in X$ with $d(m, x) \leq \epsilon$. The following result now follows from the nerve lemma 4.6.

Theorem 4.17 *Let M be a compact Riemannian manifold, and suppose that ϵ is one of the values whose existence is guaranteed by Theorem 4.16 above. Then the geometric realization of the Čech complex on an ϵ-net X with scale ϵ is homotopy equivalent to M.*

We also observe that, given a point cloud Z in a metric space E and two scales $\epsilon \leq \epsilon'$, because of the fact that $B(z, \epsilon) \subseteq B(z, \epsilon')$ there is an evident map of covered sets $(Z, \mathfrak{U}_\epsilon^{\text{Cech}}) \to (Z, \mathfrak{U}_{\epsilon'}^{\text{Cech}})$ which is the identity on Z, regarded both as the underlying set Z as well as the indexing set for the coverings $\mathfrak{U}_\epsilon^{\text{Cech}}$ and $\mathfrak{U}_{\epsilon'}^{\text{Cech}}$. We now obtain the following.

Proposition 4.18 *Given $\epsilon \leq \epsilon'$ as above, the complexes $C^{\text{Cech}}(Z, E, \epsilon)$ and $C^{\text{Cech}}(Z, E, \epsilon')$ both have Z as their vertex set, and $C^{\text{Cech}}(Z, E, \epsilon) \subseteq C^{\text{Cech}}(Z, E, \epsilon')$.*

We also note that $C^{\text{Cech}}(-, \epsilon)$ has the following functoriality property.

Proposition 4.19 *Suppose that we have point clouds $X \subseteq A$ and $Y \subseteq B$, where A and B are metric spaces, and further that $f: A \to B$ is a weak contraction, i.e. for any $a_1, a_2 \in A$ we have $d(f(a_1), f(a_2)) \leq d(a_1, a_2)$. Suppose further that f carries X into Y, so that the restriction $f|_X$ maps into Y. Then there is a unique simplicial map $C^{\text{Cech}}(f, \epsilon): C^{\text{Cech}}(X, A, \epsilon) \to C^{\text{Cech}}(Y, B, \epsilon)$ which is equal to $f|_X$ on vertices. More generally, we observe that if f has bounded expansion γ (i.e., $d_Y(f(x_1), f(x_2)) \leq \gamma d_X(x_1, x_2)$ for all $x_1, x_2 \in X$), there is an induced map from $C^{\text{Cech}}(X, A, \epsilon)$ to $C^{\text{Cech}}(Y, B, \gamma\epsilon)$.*

Proof These results follow immediately from Proposition 4.12. □

4.3.2 The Vietoris–Rips Complex

Although the Čech complex of a point cloud is theoretically more tractable than most other constructions, in many circumstances the Čech complex is not appropriate. For one thing, computing the simplices of the Čech complex for $Z \in \mathbb{R}^n$ depends on being able to decide the non-emptiness of the intersection of balls in the ambient metric space. This can often be computationally expensive and can require complicated algorithms to achieve a satisfactory performance. Furthermore, it may not be the case that the point cloud is equipped with a reasonable embedding in an ambient metric space, and although various techniques (e.g. spectral methods based on an associated graph Laplacian) can be used to produce an embedding, this adds additional complexity and potentially introduces distortion.

As such, it is often useful to work instead with a related complex which is essentially determined by its behavior on the 1-skeleton: the Vietoris–Rips complex. Once again, we fix a scale parameter $\epsilon > 0$.

Definition 4.20 The Vietoris–Rips complex of the point cloud Z at scale ϵ is the simplicial complex whose vertex set is the set of points of Z, and for which a family $\{z_0, \ldots, z_k\}$ spans a k-simplex if and only if

$$d(z_i, z_j) \leq \epsilon$$

for $0 \leq i < j \leq k$. We denote this complex by $\mathrm{VR}(X, \epsilon)$.

That is, the Vietoris–Rips complex is determined by the pairwise distances between points. A higher simplex σ is in the complex if and only if all 1-simplices that can be obtained as faces are themselves in the complex. When the Čech complex exists, the Vietoris–Rips complex is fully determined by the 1-skeleton (the edges) of the Čech complex.

The Vietoris–Rips complex has the following mapping properties.

Proposition 4.21 *Let X be any finite metric space, and suppose we are given two scale parameters $\epsilon_1 \leq \epsilon_2$. Then $\mathrm{VR}(X, \epsilon_1) \subseteq \mathrm{VR}(X, \epsilon_2)$.*

Proposition 4.22 *Let X and Y be two finite metric spaces, and suppose that $f : X \to Y$ is a weak contraction. Then there is a unique simplicial map $\mathrm{VR}(f, \epsilon) : \mathrm{VR}(X, \epsilon) \to \mathrm{VR}(Y, \epsilon)$ for any scale value $\epsilon \geq 0$. As in the case of the Čech complex, we also have that if f has bounded expansion γ then there is an induced map $\mathrm{VR}(X, \epsilon) \to \mathrm{VR}(Y, \gamma\epsilon)$.*

The Vietoris–Rips complex and the Čech complex are closely related as ϵ varies. Clearly, if a simplex $[x_1, x_2, \ldots, x_k]$ is contained in the ϵ-Čech complex then it is contained in the 2ϵ-Vietoris–Rips complex; that is, for all ϵ, we have the inclusion

$$C^{\mathrm{Cech}}(X, \epsilon) \subset \mathrm{VR}(X, 2\epsilon).$$

In the other direction, we have the following containment lemma.

Lemma 4.23 *There is an inclusion*

$$\mathrm{VR}\,(X, \epsilon/2) \subset C^{\mathrm{Cech}}(X, \epsilon).$$

Proof Suppose that $[x_1, x_2, \ldots, x_n]$ is a k-simplex in $VR(X, \epsilon/2)$. Choose a point $z \in B(x_1, \epsilon/4) \cap B(x_2, \epsilon/4)$. Then $d(z, x_1) < \epsilon/4$. Furthermore, since for all i and j we have that $d(x_i, x_j) < \epsilon/2$, the triangle inequality implies that $d(z, x_k) < 3\epsilon/4 < \epsilon$. Therefore, z is contained in $B(x_k, \epsilon)$, and so the intersection

$$\bigcap_{k=1}^{n} B(x_k, \epsilon)$$

is non-empty. □

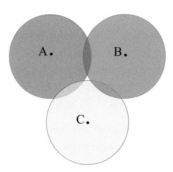

Figure 4.6 A simplex consisting of three points, A, B and C, which belongs to the Vietoris–Rips simplex but not to the Čech complex

Figure 4.6 shows a configuration of three points in the Euclidean plane which lie in the Vietoris–Rips complex at a certain threshold but not in the Čech complex for the same threshold.

In the special case where X is a subset of $\mathbb{R}^n$ and the metric on X is the restriction to the Euclidean metric, there is a useful interpretation of the Vietoris–Rips complex as the result of a certain construction on the Čech complex.

Definition 4.24 Let X denote any simplicial complex. Then by the *flag complex* of X, we mean the simplicial complex $\mathfrak{F}X$ in which a subset $\sigma \subseteq V(X)$ is a simplex of $\mathfrak{F}X$ if and only if all its subsets of cardinality 2 are 1-simplices in X. It is immediate that $X \subseteq \mathfrak{F}X$.

Proposition 4.25 *For a point cloud $X \subseteq \mathbb{R}^n$, where $\mathbb{R}^n$ is equipped with the Euclidean metric, the complex $VR(X, \epsilon)$ is exactly the complex $\mathfrak{F}C^{\text{Čech}}(X, \epsilon/2)$.*

Proof Because both complexes are determined completely by their edges, it suffices to show that the edges of the two are identical. The pair $\{x, x'\}$ is a simplex in $VR(X, \epsilon)$ if and only if $d(x, x') \leq \epsilon$. On the other hand, $\{x, x'\}$ is an edge of $\mathfrak{F}C^{\text{Čech}}(X, \epsilon/2)$ if and only if the closed balls of radius $\epsilon/2$ around x and x' intersect. This happens exactly if the distance Euclidean distance between x and x' is $\leq \epsilon$. □

Remark 4.26 Because the Vietoris–Rips complex requires much less storage and computation than the Čech complex, we can think of the Vietoris–Rips complex as the "lazy version" of the Čech complex. By applying the $\mathfrak{F}$ construction to other complex constructors, we obtain lazy versions of them. Note that the Vietoris–Rips is not the nerve of a covering in any natural way. The interest in it comes from the efficiency of its description.

4.3.3   The Alpha-Shapes Complex

We begin by discussing the Voronoi diagram. Although this can be defined in arbitrary metric spaces, for the time being we will focus on subsets of Euclidean space $X \subset \mathbb{R}^n$. Given a finite set X, there is an induced covering of $\mathbb{R}^n$ as follows.

Definition 4.27   Let X be a finite subset of $\mathbb{R}^n$. The *Voronoi cell of $x \in X$* is the set $V(x) = \{p \in \mathbb{R}^n \mid \partial(p, x) \leq \partial(p, z), z \in X\}$. That is, it is the set of points which are closer to x than any other point of X. It is clear that the family $\{V(x)\}_{x \in X}$ forms a covering of $\mathbb{R}^n$, which we refer to as the Voronoi covering attached to X and denote by $\mathfrak{U}_X^{\text{Vor}}$. The nerve of $\mathfrak{U}_X^{\text{Vor}}$ will be referred to as the *Delaunay complex* of X.

Given a subspace $W \subseteq \mathbb{R}^n$, we can consider the covering $\{V(x) \cap W\}_{x \in X}$ of W. The nerve of this covering is referred to as the *restricted Delaunay complex*, and we denote it by $\text{Del}_W X$.

Proposition 4.28   *Provided that all the restricted Voronoi cells are contractible, the restricted Delaunay complex of X is weakly equivalent to X.*

Proof   This follows from the nerve lemma (Theorem 4.6).   □

For any $\epsilon \geq 0$, we let $X(\epsilon)$ denote the union of all open balls of radius ϵ centered at points of X.

Definition 4.29   *Given a finite subset $X \subseteq \mathbb{R}^n$ and $\epsilon \geq 0$, we define the α-shape complex of X with scale parameter ϵ to be the restricted Delaunay complex $\text{Del}_{X(\epsilon)} X$.* We will denote it by $C^\alpha(X, \epsilon)$.

Proposition 4.30   *If $\epsilon \leq \epsilon'$ then there is an inclusion*

$$C^\alpha(X, \epsilon) \hookrightarrow C^\alpha(X, \epsilon').$$

Proof   Since it is immediate that $X(\epsilon) \subseteq X(\epsilon')$, this follows directly from Proposition 4.12.   □

It is sometimes useful to replace $X(\epsilon)$ with a space defined using varying radii of balls for points of X. It is called the *weighted Voronoi diagram* associated with a set of balls $B(x_i, r_i)$. We refer the reader to Edelsbrunner et al. (1983) for more information on this construction.

4.3.4   Witness Complexes

The construction of the Voronoi cells for a finite subset $Z \subseteq \mathbb{R}^n$ can be carried out formally for a finite subset of any metric space. Nothing in the definition uses any properties of $\mathbb{R}^n$. In particular, it can be employed where the ambient space X is itself a finite metric space. Of course, in this case the geometry of the cells is much more difficult to understand, and they do not reflect the geometry of the data well.

Example 4.31 Let X be the subspace of $\mathbb{R}$ consisting of all the numbers of the form $k/100$ for $k \in \{n \in \mathbb{Z} | n \neq 0\}$. Using X as a replacement for $\mathbb{R}$, we let $Z \subseteq X$ denote the subset $\{\pm 1\}$. The Voronoi cells for the inclusion $Z \subseteq \mathbb{R}$ would be the two half-infinite intervals $(-\infty, 0]$ and $[0, +\infty)$, which overlap at the origin. In this case, the Delaunay complex becomes the standard 1-simplex, which is arguably a reasonable representation of $\mathbb{R}$. On the other hand, if we produce the analogue of the Delaunay complex for the inclusion $Z \hookrightarrow X$, we find that the two analogues to the Voronoi cells do not intersect, because the origin is not a part of x. However, we would expect that a construction of this type should give results for the two situations that are comparable, since X is a relatively dense subset of $\mathbb{R}$. The moral is that, when working with data, one should be wary of exact equalities.

Despite these remarks, one can certainly form the restricted Delaunay complex for an inclusion of a finite set Z into any finite metric space, and we can ask whether there are ways to enlarge the Voronoi cells "a little bit", to allow the complex to reflect the geometry of the spaces better.

Definition 4.32 Let X be any metric space, and let $Z \subseteq X$ be any finite subset. For any $z \in Z$, let the ϵ-*Voronoi cell for z in x* be the set

$$\{x \in X | d(x, z) \leq d(x, \zeta) + \epsilon \text{ for all } \zeta \in Z\}.$$

We define the ϵ-*Voronoi diagram for X with landmark set Z and threshold ϵ* to be the nerve of the covering of X by the ϵ-Voronoi cells for X relative to Z. We will denote this simplicial complex by $C^V(X, Z, \epsilon)$. There is also a "lazy" version of this complex whose vertices and edges are identical to those of $C^V(X, Z, \epsilon)$ and where a higher simplex is present in the complex if and only if all its edges are thus present. It bears the same relationship to $C^V(X, Z, \epsilon)$ as the VR-complex bears to C^{Cech}.

Remark 4.33 In Example 4.31 above, if $\epsilon = \frac{1}{100}$, we find that the two ϵ-Voronoi cells do intersect. In this case, $C^V(X, Z, \epsilon)$ is the interval complex with two vertices and a 1-simplex connecting the two.

We observe that if $\epsilon \leq \epsilon'$ then the ϵ-Voronoi cell for $z \in Z$ is contained in the ϵ'-Voronoi cell for z. What this means is that the following analogue to Propositions 4.18, 4.21, and 4.30 holds.

Proposition 4.34 *Suppose we are given a finite subset $Z \subseteq X$, where X is a metric space, and suppose that $0 \leq \epsilon \leq \epsilon'$; then the complexes $C^V(X, Z, \epsilon)$ and $C^V(X, Z, \epsilon')$ both have Z as their vertex set, and*

$$C^V(X, Z, \epsilon) \subseteq C^V(X, Z, \epsilon').$$

Remark 4.35 In the case of the alpha-shapes complex, we used the embedding of a point cloud X in $\mathbb{R}^n$ to generate a union of balls whose topology we expect to reflect the geometric properties of X. We are *not* interested in the topology of the ambient space $\mathbb{R}^n$, since of course it is known to us already. In the case of ϵ-Voronoi diagrams, we typically consider situations where both X and $Z \subseteq X$ are finite metric spaces, and Z is considered to be a "landmark set" for building a simplicial complex and therefore a topological space which we expect will reflect the geometric features of X. We will therefore use the letter $\mathcal{L}$ for the subset Z.

The vertex set attached to the complex $C^V(X, \mathcal{L}, \epsilon)$ is in one-to-one correspondence with $\mathcal{L}$ rather than with X. What this means is that, as a simplicial complex, it will be much smaller than C^{Cech}, VR, and C^α. Moreover, one is free to select the number of landmarks with which one can afford to compute. This is quite important, since a worst-case number of simplices for C^{Cech} and VR gives the number of simplices at $O(2^n)$. The alpha-shapes complex C^α provides a significant sparsification of the set of higher simplices but it requires an embedding of the data set in $\mathbb{R}^n$, which is not always obtainable in a rational way. Also, if the dimension n is chosen too large, the complexity of computing with the Voronoi diagram is prohibitive (see Klee 1980).

For later use, we will also define a complex which depends on two distinct landmark sets Z and W and which we will refer to as the *bivariate ϵ-Voronoi diagram for X relative to Z and W*. Its vertex set is the subset $\mathcal{V}(Z, W)$ of $Z \times W$ consisting of pairs (z, w) for which the Voronoi cells attached to z and w (relative to the landmark sets Z and W) have at least one point in common and the set $\{(z_0, w_0), \ldots, (z_k, w_k)\}$ spans a simplex if and only if the intersection

$$(V^Z(z_0) \times V^W(w_0)) \cap \cdots \cap (V^Z(z_k) \times V^W(w_k))$$

is non-empty, where $V^Z(-)$ and $V^W(-)$ denote the Voronoi cells computed for the landmark sets Z and W, respectively. It is easy to check that the vertex maps $\mathcal{V}(Z, W) \to Z$ and $\mathcal{V}(Z, W) \to W$ given by projections on Z and W, respectively, induce maps of simplicial complexes.

At the present time the complex C^V has not yet been implemented. However, variants of it that accelerate the appearance of homology as ϵ varies were constructed in de Silva & Carlsson (2004) and were called *witness complexes*. We discuss them now.

To motivate the discussion, let us first note that a family $\{l_0, \ldots, l_k\}$ is a k-simplex in $C^V(X, \mathcal{L}, 0)$ if and only if there exists a point $x \in X$ such that

$$d(l_i, x) = \min_{l \in \mathcal{L}} d(l, x).$$

In particular, $d(l_i, x) = d(l_j, x)$ for all $l_i, l_j \in \mathcal{L}$. We will reformulate this condition in such a way that there is a natural way to relax it. Given a finite *multiset* (or *bag*) A of real numbers (see Hein 2003, Section 1.2.4, or Monro 1987 for a discussion of multisets), we say that a submultiset $S \subseteq A$ is *minimizing in A* if, for all $a \in A - S$, we have that $a \leq s$ for all $s \in S$. In the case of an edge $\{l_0, l_1\}$, we can write the membership condition for simplices of $C^V(X, \mathcal{L}, 0)$ as equivalent to the existence of a point x satisfying the following two conditions:

1. $d(l_0, x) = d(l_1, x)$;
2. the set $\{d(l_0, x), d(l_1, x)\}$ is minimizing in the set $\{d(l, x)\}_{l \in \mathcal{L}}$.

We can relax the membership condition by removing the first condition, i.e. that l_0 and l_1 be equidistant to x. There is a natural extension of condition 2 above to higher-dimensional simplices, which, given a potential simplex $\sigma = \{l_0, \ldots, l_k\}$, would require that there is an $x \in X$ such that the set $\{d(l_0, x), \ldots, d(l_k, x)\}$ is minimizing in the set $\{d(l, x)\}_{l \in \mathcal{L}}$. Unfortunately, this condition is not necessarily inherited by the faces of σ, so instead we impose the condition that, for every face $\tau = \{l_{i_0}, \ldots, l_{i_s}\}$ of σ, there is an element $x_\tau \in X$, called a *witness for* τ, such that the set $\{d(l_{i_0}, x_\tau), \ldots, d(l_{i_s}, x_\tau)\}$ is minimizing in the set $\{d(l, x_\tau)\}_{l \in \mathcal{L}}$. More generally, we introduce a notion of witness that is dependent on a parameter ϵ. As before, let A be a finite multiset of real numbers, let $S \subseteq A$ be a submultiset, and let $\overline{S} = A - S$. If $\epsilon \geq 0$, we say that S is ϵ-minimizing if S is minimizing in the multiset $S \cup \epsilon + \overline{S}$, where, for any multiset R of real numbers, $\epsilon + R$ is the multiset $\{\epsilon + r | r \in R\}$. Given a submultiset $\sigma = \{l_{i_0}, \ldots, l_{i_s}\} \subseteq \mathcal{L}$, an element $x \in X$ is said to be an ϵ-witness for σ if $\{d(l_{i_0}, x) \ldots, d(l_{i_s}, x)\}$ is ϵ-minimizing in $\{d(l, x)\}_{l \in \mathcal{L}}$. Note that if $\epsilon \leq \epsilon'$, and x is an ϵ-witness for a simplex σ, then it is also an ϵ'-witness for σ. We define the witness complex $W_\infty(X, \mathcal{L}, \epsilon)$ as the simplicial complex with vertex set $\mathcal{L}$ for which a non-empty subset $\sigma \subseteq \mathcal{L}$ is a simplex if and only if σ and all its faces have ϵ-witnesses. We denote it by $W_\infty(X, \mathcal{L}, \epsilon)$ and now describe the relationship between C^V and W_∞.

Proposition 4.36 *There is an inclusion of simplicial complexes* $C^V(X, \mathcal{L}, \epsilon) \subseteq W_\infty(X, \mathcal{L}, \epsilon)$.

Proof A subset $\{l_0, \ldots, l_k\}$ spans a k-simplex of $C^V(X, \mathcal{L}, \epsilon)$ if and only if there exists an element $x \in X$ such that $d(l_i, x) \leq d(l, x) + \epsilon$ for all i and $l \in \mathcal{L}$. Let A denote the multiset $\{d(l, x)\}_{l \in \mathcal{L}}$. Then it clearly follows that the set $S = \{d(l_0, x), \ldots, d(l_k, x)\}$ is minimizing in the multiset $S \cup \epsilon + (A - S)$, so x is an ϵ-witness for the set $\{l_0, \ldots, l_k\}$. $\square$

By applying the construction $\mathfrak{F}$, we can obtain a lazy version $W_\infty^{\text{lazy}} = \mathfrak{F} W_\infty$, and consequently a natural inclusion $C_{\text{lazy}}^V = \mathfrak{F}(C^V) \rightarrow \mathfrak{F}(W_\infty) = W_\infty^{\text{lazy}}$.

Remark 4.37 Some of the constructions we have given above were introduced in de Silva & Carlsson (2004). The complex $W_\infty(X, \mathcal{L}, 0)$ is identical to the complex $W_\infty(D)$ introduced there. The input D is the $N \times L$ matrix of distances from all landmark points to all points of X, where $N = \#(X)$ and $L = \#(\mathcal{L})$, and so contains all the information concerning X and $\mathcal{L}$ required to perform the construction. Similarly, the construction $W_\infty^{\text{lazy}}(X, \mathcal{L}, \epsilon)$ is identical to the construction $W(D, \epsilon, 2)$ described in de Silva & Carlsson (2004). There is a family of complexes $W(D, \epsilon, \nu)$ which are all different versions of lazy constructions, and the $\nu = 2$ case of the construction is identified with our W_∞^{lazy}. We choose not to introduce the complexes for the other values of ν, since they are not often used. However, the case $\nu = 0$ is closely related to the Vietoris–Rips complex on $\mathcal{L}$, and $W(D, \epsilon, 1)$ bears a relationship to the complex $C_{\text{lazy}}^V(X, \mathcal{L}, \epsilon)$.

The goal of these constructions is to make persistent homology, which we will be introducing in Section 4.5.1, more computable by permitting calculation with relatively small complexes. Both the Čech and Vietoris–Rips complexes have vertex sets of size N, where N is the cardinality of the point cloud, and this means that the size of the set of k-simplices can be expected to grow as $O(N^{k+1})$. Since point clouds of interest often contain thousands or millions of points, the calculation is often intractable. What is permitted by Voronoi-based constructions is to compute with a much smaller set, namely the landmark set $\mathcal{L}$. It turns out that there is an additional benefit, namely that Voronoi-based constructions often remove much of the noise in the persistent homology barcodes. The notion of noise will be interpreted when we introduce persistent homology in Section 4.5.1. This noise-removal property is most strongly in evidence when we are using the complex W_∞^{lazy}. In addition, use of the Voronoi-based constructions means that we are not even required to compute all the distances in the point cloud, only the distances from landmark points to arbitrary points of X, i.e. $N \times L$ values rather than N^2.

Of course, it is important to have good methods for selecting landmarks. One would clearly like them to be well distributed within X. In de Silva & Carlsson (2004) it is recommended that one should use either a random selection of landmark points from a point cloud or the *maxmin* method.

The maxmin method proceeds according to the following algorithm.

1. Pick some initial point ℓ_1 at random.
2. Having picked points $\ell_1, \ldots, \ell_{n-1}$, pick the next landmark point ℓ_n so as to maximize

$$\min\{d(\ell_1, \ell_n), \ldots, d(\ell_{n-1}, \ell_n)\}.$$

Repeat until sufficiently many points have been picked.

Maxmin picks more evenly spaced landmark points but tends to pick out extremal values as landmarks.

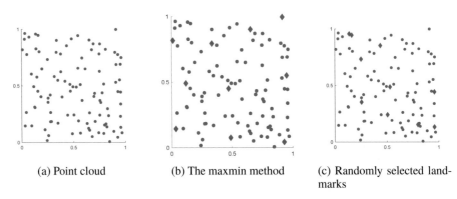

(a) Point cloud   (b) The maxmin method   (c) Randomly selected landmarks

Figure 4.7 Landmark selection on 100 points sampled uniformly from the unit square. The data points are round blue dots, while the selected landmarks are marked as red diamonds.

4.3.5   Mapper

One way to use simplicial complex models of point cloud data is to work with them directly to interrogate the data set in various ways. The idea is as follows. Suppose that we have a point cloud X and a covering $\mathfrak{U}$ of X, so that we can construct the nerve $N\mathfrak{U}$ as a model for the data. If $N\mathfrak{U}$ is sufficiently low-dimensional (as a simplicial complex), it can be embedded in $\mathbb{R}^2$ or $\mathbb{R}^3$. This data can be used to permit a user to interact with X in various ways, using the fact that sets of vertices in $N\mathcal{U}$ correspond to subsets of X. For any set of vertices V of $N\mathcal{U}$, we write X_V for the corresponding subset of X.

1. One can use standard graph layout algorithms to view the simplicial complex $N\mathcal{U}$, thus providing a visualization tool for point clouds. This visualization is most effective when the complex is low-dimensional.

2. Functions on X can be used to construct functions on the vertex set of $N\mathcal{U}$ using an averaging procedure. These functions can in turn be represented by colorings of the set of vertices of $N\mathcal{U}$. In this way one can, for example, locate "hot spots" for values of interest, such as those corresponding to survival or revenue, within X.

3. One can build user interfaces which allow one to select regions within $N\mathcal{U}$ (and therefore subsets of X) by point-and-click methods, as in Adobe Illustrator or Photoshop. This ability to select localized subsets within X is extremely useful, and enables many ways of working with the data. For example, a selected group can be treated as a data set in its own right, and analyzed to obtain a more local understanding of the data set. This often gives more fine-grained information about X.

4. One can select families of subgroups by hand to obtain segmentations of X.

5. Given a subset $X_0 \subseteq X$, we can define a function on the set of vertices v of $N\mathcal{U}$ to be the fraction of elements of $X_{\{v\}}$ that lie in X_0. In this way one can understand if and where a chosen group is localized in X.

6. One can obtain "explanations" characterizing subgroups that have been selected or defined. This is done by considering all the variables, and for each computing a Kolmogorov–Smirnov (KS) score comparing the distribution of the variable on the subgroup to that on the entire set. The variables can then be ordered by this score, so that one obtains the variables with highest KS score first in the list.

7. If we have two distinct coverings $\mathcal{U}$ and $\mathcal{U}'$ of X, it is often useful to produce layouts of both nerve complexes $N\mathcal{U}$ and $N\mathcal{U}'$ simultaneously. This permits the selection of a group V of vertices in $N\mathcal{U}$, then for each vertex v of $N\mathcal{U}'$ computing the percentage of points of $X_{\{v\}}$ which lie in the subcollection X_V, and finally coloring the vertex set of $N\mathcal{U}'$. This procedure gives a very good way to compare models.

8. If we suppose that our data is given as the rows of a rectangular matrix, it is often useful to build a topological model not on the data points (rows) of the matrix, but rather on the columns (features) of the matrix. Each data point of the original matrix can now be regarded as a function on the set of features, and, as in point 2 above, we construct the corresponding function on the vertices of the topological model of the set of features. This process gives a useful way to understand data sets with many features directly. There can be situations where the number of rows is so small that building a useful topological model of them is not possible but where the number of features is sufficiently large to support such a model, and this method allows one to study directly a small but high-dimensional point cloud. By a further averaging procedure, one can obtain functions based on subsets of the X as well.

The existence of this kind of functionality means that it is very useful to construct coverings of point clouds X in interesting ways. One such construction is called *Mapper* and was introduced in Singh et al. (2007). It is based on the *Reeb graph* construction (see Reeb 1946), which we outline.

Let $f \colon X \to \mathbb{R}$ be a continuous map, where X is a topological space. Then we define an equivalence relation $\simeq$ on the space X as follows. Given $x, x' \in X$, we have that $x \simeq x'$ if (a) $f(x) = f(x')$ *and* (b) x and x' belong to the same connected component of $f^{-1}(f(x))$. The Reeb graph $R(f)$ is defined to be the quotient $X/\simeq$. It is equipped with a map to the real line that sends an equivalence class to the value of f on any of its members.

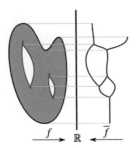

Figure 4.8 A Reeb graph. Reproduced from Silva et al. (2016) by permission from Springer-Nature, © 2016.

It is of course possible to replace the target space $\mathbb{R}$ by any space B, and perform an analogous construction. This produces a space which maps to B and in which the inverse images of points are discrete spaces. Given a continuous map $f \colon X \to B$, we will refer to this construction as the *Reeb complex* of f, and denote it by $\mathfrak{R}(f)$.

In order to extend the construction to point clouds, we will need to understand how to construct coverings of Reeb complexes.

Definition 4.38 Let $f \colon X \to B$ be a map, and let $\mathfrak{U} = \{U_\alpha\}_{\alpha \in A}$ be a finite open covering of B. For each $\alpha \in A$, we may consider the collection of connected components of $f^{-1}(U_\alpha)$ given the subspace topology. They are open sets in X, and we will denote the covering consisting of all of them by $\mathfrak{U}^{\mathrm{Reeb}}(f)$.

Remark 4.39 The coverings $\mathfrak{U}^{\mathrm{Reeb}}(f)$ are in general infinite, but there are simple finiteness conditions on the space X which will guarantee that they are finite. Since the discussion is only intended to provide motivation for the the point cloud version, we will not worry about this technical point but will allow ourselves to discuss the nerves of these coverings.

The nerve of $\mathfrak{U}^{\mathrm{Reeb}}(f)$ is equipped with a natural map of simplicial complexes $N\mathfrak{U}^{\mathrm{Reeb}}(f) \to N\mathfrak{U}$.

Now suppose that we have a point cloud X with a reference map f to a reference space B and that B is equipped with a finite open covering $\mathfrak{U} = \{U_\alpha\}_{\alpha \in A}$. Then we have a covering of X by the sets $f^{-1}(U_\alpha)$. In Definition 4.38, we further decomposed these subsets into their connected components. Since we are dealing with point clouds, we replace the construction of the connected components by the partition produced by a clustering algorithm, which may be chosen by the user of the method. By the Mapper complex for f and the given clustering algorithm, we will mean the nerve of the covering of X by all these clusters. Note that we do *not* cluster the entire set X but only the subsets $f^{-1}(U_\alpha)$. The Mapper complex is equipped with a map to $N\mathfrak{U}$.

In the actual applications of this method, the space B is usually a product of a small number of intervals (1, 2, or 3), and the maps to the various intervals can be chosen in different ways:

- using individual features in the point cloud X:
- using statistically meaningful quantities, such as the output of a density estimator or centrality measure;
- using the output of an algebraic machine-learning construction such as principal component analysis or multidimensional scaling.

In the case where B is a closed interval, the coverings chosen are typically coverings by families of open intervals of equal length and for which we permit the overlaps to be either empty or a fixed percentage of the length of the interval. In the case of products of intervals, we choose such coverings for each interval and then form products of them. We thus obtain coverings of the product of intervals by rectangles or higher-dimensional analogues. Finally, in Singh et al. (2007), the clustering method used was single-linkage hierarchical clustering with a particular heuristic for the choice of threshold parameter. We refer to Singh et al. (2007) for the specific heuristic used.

This method has seen applications in many different situations. See for example Nicolau et al. (2011), Torres et al. (2016), Saggar et al. (2018), Duponchel (2018), Li et al. (2015), and Offroy & Duponchel (2016). We discuss one example in detail in Section 6.11.

4.4 Persistence

Under favorable conditions, the complexes introduced in the previous sections approximate the homotopy type of the "underlying" space, provided that there are sufficiently many points and that ϵ is large enough but smaller than the feature scale. Unfortunately, when working with real data the correct choice of value for ϵ is fundamentally unknowable. While it is possible to attempt to infer reasonable ranges for ϵ from the data, the idea that we pursue in this section is that instead it makes sense to *keep track of all ϵ values at once*. For each ϵ, we obtain a simplicial complex (as described in the preceding section) and these complexes fit together.

4.4.1 Filtered Simplicial Complexes

Let Z be an abstract simplicial complex. We define a filtration on Z as a sequence of simplicial complexes

$$\emptyset = Z^0 \subseteq Z^1 \subseteq \cdots \subseteq \cdots Z.$$

We require that Z^i be a subcomplex of Z^{i+1} and each Z^i a subcomplex of Z. Since we are primarily interested in finite complexes, we will typically make the assumption that the filtration stabilizes after some N, i.e. that $Z^n = Z$ for $n \geq N$. We will refer to Z with a given filtration $\{Z^i\}$ as a filtered simplicial complex.

Given filtered simplicial complexes $Z, \{Z^i\}$ and $Y, \{Y^i\}$, a map of the filtered simplicial complexes is a simplicial map $f : Y \to Z$ such that $f(Z^i) \subseteq Y^i$; in this way we get a family of simplicial maps $f^i : Z^i \to Y^i$ by restriction.

Example 4.40 Let K be a simplicial complex. Then there is a natural filtered simplicial complex associated with K given by the skeletal filtration $K^i = K_i$. A simplicial map $K \to K'$ induces maps of skeletally filtered complexes. This filtration stabilizes if and only if K is a finite-dimensional complex.

Here, we use the notation K_i for the i-skeleton of K, i.e. the subcomplex that contains all simplices in K of dimension $\leq i$.

Example 4.41 Let K_ϵ be a family of simplicial complexes, determined by a parameter ϵ. Suppose that K_ϵ is functorial in the sense that if $\epsilon < \epsilon'$ then $K_\epsilon \subseteq K_{\epsilon'}$, and further suppose that the complex changes finitely many times.

Then, for some sufficiently dense set $\{t_1, \ldots t_n\}$ of parameter values, a filtration of K_{t_n} is given by the sequence $\emptyset \subseteq K_{t_1} \subseteq \cdots \subseteq K_{t_n}$.

All the complexes defined in Section 4.3 give rise to filtered simplicial complexes in this way.

Associated with a filtered simplicial complex, we obtained filtered algebraic structures from the standard constructions of algebraic topology. Recall from Section 3.3 that, associated the simplicial complex K and a ring R, we obtain a chain complex $C_*(K; R)$:

$$\cdots \longrightarrow C_{k+1}(K; R) \xrightarrow{\partial_{k+1}} C_k(K; R) \xrightarrow{\partial_k} C_{k-1}(K; R) \longrightarrow \cdots,$$

where the connecting maps are the boundary maps given in Definition 3.118. For each k, there are distinguished subgroups of $C_k(K; R)$, the cycles $Z_k(K; R)$ and the boundaries $B_k(K; R)$. These are related by the system of inclusions

$$B_k(K; R) \subseteq Z_k(K; R) \subseteq C_k(K; R)$$

and, since $\partial_k \partial_{k+1} = 0$, we can define the homology $H_k(K; R) = Z_k(K; R)/B_k(K; R)$.

4.4.2   Euler Characteristic Curve

The Euler characteristic provided one of the first homological summaries of these filtered complexes used for topological data analysis.

Definition 4.42 The Euler characteristic $\chi(X)$ of a simplicial complex is the alternating sum of its counts of simplices by dimension:

$$\chi(X) = \sum_{i=0}^{d} (-1)^i \dim C_i(X).$$

The name comes from its use in the Euler polyhedral formula: the Euler characteristic of a polyhedron P with V vertices, E edges and F faces is $\chi(P) = V - E + F$. We can note that, for the platonic solids and for this torus made up of 16 quadrilaterals,

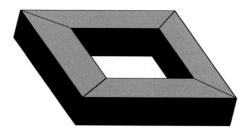

we get the Euler characteristics shown at the bottom of the following table:

Shape	Tetrahedron	Cube	Octahedron	Dodecahedron	Icosahedron	Torus
V	4	8	6	20	12	16
E	6	12	12	30	30	32
F	4	6	8	12	20	16
χ	2	2	2	2	2	0

Remarkably, this cell-based count is a homology invariant: for simplicial complexes with isomorphic homology groups, the Euler characteristic will be the same. One way to prove this leads straight into the first algorithms for computing persistent homology (which we will introduce next, in Section 4.5.1).

Theorem 4.43

$$\sum_{i=0}^{d}(-1)^i \dim C_i(X) = \sum_{i=0}^{d}(-1)^i \dim H_i(X).$$

Proof Consider some total ordering $\sigma_1, \sigma_2, \ldots$ of the simplices of the complex X in such a way that all simplices in $\partial\sigma$ appear before σ itself. By adding one simplex at a time we can show that any changes in homology groups will match the changes in chain groups.

We traverse the simplices in order, and maintain a basis of $C_*(X)$ split into three subspaces as follows: $C_*(X) = B_*(X) \oplus Z_*(X) \oplus N_*(X)$, where $B_*(X) = \operatorname{Im}\partial$ and $B_*(X) \oplus Z_*(X) = \operatorname{Ker}\partial$. By maintaining this split, $Z_*(X)$ will have basis elements precisely corresponding to a selection of representative cycles of each non-zero homology class in X.

For each basis element, we track the oldest simplex in its support, and we maintain basis choices such that

1. Each basis element in $N_*(X)$ is monomial: it corresponds to just a single simplex.
2. No two basis elements in $B_*(X) \oplus Z_*(X)$ share the newest simplex in their support.

We can maintain this because if $z = \sigma + z'$ and $w = \sigma + w'$ then $z - w = z' - w'$ is a new basis element, which may replace z without a significant change in basis choice. Clearly we can recover z from w and $z - w$ and, since $\partial z = 0$ and $\partial w = 0$, we also have $\partial(z - w) = 0$ since the boundary operator is linear.

If $z, w \notin B_*(X)$ then $z - w$ cannot be in $B_*(X)$ either: if it were, then, for some u, we would have $\partial u = z - w$ and thus $[z - w] = 0$ in the homology group. Thus $[z] = [w]$. Since the basis for $Z_*(X)$ chooses exactly one cycle for each homology class, $z - w \in B_*(X)$ would imply that $z = w$ since otherwise we would have two cycles representing the same homology class.

The same argument gives that if $z \notin B_*(X)$ and $w \in B_*(X)$ then $z - w \notin B_*(X)$ since otherwise $[z] = [z - w] = 0$, which would mean that at the outset z was not a representative of a non-trivial homology class.

As we add a new simplex σ to produce $X \cup \sigma$, the boundary $\partial_{X \cup \sigma} \sigma$ will be in $\mathrm{Ker}\, \partial_X$, since the kernel is contained in the image and nothing has changed in dimensions lower than $\dim \sigma$. So we can express $\partial_{X \cup \sigma} \sigma$ in the chosen basis. There are two possibilities.

- If the basis elements used contain at least one basis element from $Z_*(X)$ then the boundary of the simplices σ is a cycle in X. This cycle now becomes a boundary. If there are several choices available, as our convention we may pick a cycle basis element whose newest simplex is the oldest among the available choices. This basis element can be moved from $Z_*(X)$ to $B_*(X \cup \sigma)$.

- If the basis elements used contain only basis elements from $B_*(X)$ then the boundary of σ is a boundary in X. If this is the case, there is some τ such that $\partial_{X \cup \sigma} \sigma = \partial_X \tau$. Hence $\partial_{X \cup \sigma}(\sigma - \tau) = 0$, so $\sigma - \tau \in \mathrm{Ker}\, \partial_{X \cup \sigma}$ and we can add $\sigma - \tau$ as a basis element to $Z_*(X)$.

Now consider the changes in homology as we add σ: adding σ either reduces the number of $(\dim \sigma - 1)$-dimensional basis elements in $Z_*(X)$ or increases the number of $(\dim \sigma)$-dimensional basis elements in $Z_*(X)$. Either way, adding σ contributes to $\chi(X \cup \sigma)$ by $(-1)^{\dim \sigma}$.   □

We can see this correspondence mirrored if we repeat the above calculations for our six shapes but with Betti numbers instead of simplex counts.

Shape	Tetrahedron	Cube	Octahedron	Dodecahedron	Icosahedron	Torus
β_0	1	1	1	1	1	1
β_1	0	0	0	0	0	2
β_2	1	1	1	1	1	1
χ	2	2	2	2	2	0

All the platonic solids are homeomorphic to the sphere, and all share the same homology groups and thus the same Euler characteristic.

These Euler characteristics, and in particular the incremental approach outlined in the proof of Theorem 4.43, can be used to produce a functional statistic of point clouds. First we create a filtered simplicial complex, and then we traverse this complex in an order consistent both with the filtration and the forcing boundaries of a simplex being present before the simplex is added. The Euler characteristics of each filtration step in turn produces a sequence of numbers, tracing out a curve that stays approximately the same as long as the point clouds are approximately similar.

4.5   The Algebra of Persistence Vector Spaces

We first define persistence vector spaces.

Definition 4.44   Let $\Bbbk$ be any field. Then by a *persistence vector space* over $\Bbbk$, we will mean a family of $\Bbbk$-vector spaces $\{V_r\}_{r \in [0,+\infty)}$, together with linear transformations $L_V(r, r') : V_r \to V_{r'}$ whenever $r \le r'$, such that $L_V(r', r'')L_V(r, r') = L_V(r, r'')$ for all $r \le r' \le r''$. A *linear transformation f* of persistence vector spaces over $\Bbbk$ from $\{V_r\}$ to $\{W_r\}$ is a family of linear transformations $f_r : V_r \to W_r$ such that for all $r \le r'$ all the diagrams

$$
\begin{array}{ccc}
V_r & \xrightarrow{L_V(r,r')} & V_{r'} \\
f_r \downarrow & & \downarrow f_{r'} \\
W_r & \xrightarrow{L_W(r,r')} & W_{r'}
\end{array}
$$

commute in the sense that

$$f_{r'} \circ L_V(r, r') = L_W(r, r') \circ f_r.$$

A linear transformation is an *isomorphism* if it admits a two-sided inverse. A *sub-persistence vector space* of $\{V_r\}$ is a choice of $\Bbbk$-subspaces $U_r \subseteq V_r$, for all $r \in [0, +\infty)$, such that $L_V(r, r')(U_r) \subseteq U_{r'}$ for all $r \le r'$. If $f : \{V_r\} \to \{W_r\}$ is a linear transformation then the *image* of f, denoted by $im(f)$, is the sub-persistence vector space $\{im(f_r)\}$.

The notion of a quotient space also extends to persistence vector spaces. If $\{U_r\} \subseteq \{V_r\}$ is a sub-persistence vector space then we can form the persistence vector space $\{V_r/U_r\}$, where $L_{V/U}(r, r')$ is the linear transformation from V_r/U_r to $V_{r'}/U_{r'}$ given by sending the equivalence class $[v]$ to the equivalence class $[L_V(r, r')(v)]$ for any $v \in V_r$.

We will also want to extend the notion of the free vector space on a set. Let X be any set, equipped with a function $\rho : X \to [0, +\infty)$. We will refer to such a pair (X, ρ) as an $\mathbb{R}_+$-*filtered set*. Then by the *free persistence vector space* on the pair (X, ρ), we mean the persistence vector space $\{W_r\}$, with $W_r \subseteq V(X)$ equal to the $\Bbbk$-linear span of the set $X[r] \subseteq X$ defined by $X[r] = \{x \in X | \rho(x) \le r\}$. Note that $X[r] \subseteq X[r']$ when $r \le r'$, so there is an inclusion $W_r \subseteq W_{r'}$. The following is a simple observation.

Proposition 4.45   *A linear combination $\sum_x a_x x \in V(X)$ lies in W_r if and only if $a_x = 0$ for all x with $\rho(x) > r$.*

We will write $\{V(X, \rho)_r\}$ for this persistence vector space. We say a persistence vector space is *free* if it is isomorphic to one of the form $V(X, \rho)$ for some (X, ρ), and we say it is *finitely generated* if X can be taken to be finite.

Definition 4.46   A persistence vector space is *finitely presented* if it is isomorphic to a persistence vector space of the form $\{W_r\}/im(f)$ for some linear transformation $f : \{V_r\} \to \{W_r\}$ between finitely generated free persistence vector spaces $\{V_r\}$ and $\{W_r\}$; here $im(f)$ is the image vector space of the linear transformation f.

The choice of bases for vector spaces V and W allows us to represent linear transformations from V to W by matrices. We will now show that there is a similar representation for linear transformations between free persistence vector spaces. For any pair (X, Y) of finite sets and field $\Bbbk$, an (X, Y)-*matrix* is an array $[a_{xy}]$ of elements a_{xy} of $\Bbbk$. We write $r(x)$ for the row corresponding to $x \in X$, and $c(y)$ for the column corresponding to y. For any finitely generated free persistence vector space $\{V_r\} = \{V(X, \rho)_r\}$, we observe that $V(X, \rho)_r = V(X)$ for r sufficiently large, since X is finite. Therefore, for any linear transformation $f \colon \{V(Y, \sigma)_r\} \to \{V(X, \rho)_r\}$ of finitely generated free persistence vector spaces, f gives a linear transformation $f_\infty \colon V(Y) \to V(X)$ between finite-dimensional vector spaces over $\Bbbk$, and using the bases $\{\varphi_x\}_{x \in X}$ of $V(X)$ and $\{\varphi_y\}_{y \in Y}$ of $V(Y)$ determines an (X, Y)-matrix $A(f) = [a_{xy}]$ with entries in $\Bbbk$. Note that in order to obtain our usual notion of a matrix as a rectangular array we are required to impose total orderings on X and Y, but none of the matrix manipulations that we will perform will require them.

Proposition 4.47 *The (X, Y)-matrix $A(f)$ has the property that $a_{xy} = 0$ whenever $\rho(x) > \sigma(y)$. Any (X, Y)-matrix A satisfying these conditions uniquely determines a linear transformation of persistence vector spaces*

$$f_A \colon \{V(Y, \sigma)_r\} \to \{V(X, \rho)_r\},$$

and the correspondences $f \to A(f)$ and $A \to f_A$ are inverses to each other.

Proof   The basis vector y lies in $V(Y, \sigma)_{\sigma(y)}$. On the other hand,

$$f(\varphi_y) = \sum_{x \in X} a_{xy} \varphi_x.$$

By Proposition 4.45, $\sum_{x \in X} a_{xy} \varphi_x$ lies in $V(X, \rho)_{\sigma(y)}$ only if all coefficients a_{xy}, for $\rho(x) > \sigma(y)$, are zero.   □

When we are given a pair of $\mathbb{R}_+$-filtered finite sets (X, ρ) and (Y, σ), we will call an (X, Y)-matrix satisfying the conditions of Proposition 4.47 (ρ, σ)-*adapted*.

Suppose now that we are given (X, ρ) and (Y, σ), with ρ and σ both $[0, +\infty)$-valued functions on X and Y, respectively. Then any matrix $A = [a_{xy}]$ satisfying the conditions of Proposition 4.47 determines a persistence vector space via the correspondence

$$A \overset{\theta}{\longmapsto} V(Y, \sigma)/im(f_A).$$

We have the following facts about this construction.

Proposition 4.48   *For any A as described above, $\theta(A)$ is a finitely presented persistence vector space. Moreover, any finitely presented persistence vector space is isomorphic to one of the form $\theta(A)$ for some such matrix A.*

Proof   This follows directly from the correspondence between matrices and linear transformations given in Proposition 4.47.   □

Proposition 4.49   *Let (X, ρ) be an $\mathbb{R}_+$-filtered set. Then, under the correspondence between matrices and linear transformations, the automorphisms of $V(X, \rho)$ can be identified with the group of all invertible (ρ, ρ)-adapted (X, X)-matrices.*

We now have the following sufficient criterion for $\theta(A)$ to be equal to $\theta(A')$; it is entirely analogous to Proposition 3.82.

Proposition 4.50   *Let (X, ρ) and (Y, σ) be $\mathbb{R}_+$ filtered sets, and let A be a (ρ, σ)-adapted (X, Y)-matrix. Let B and C be respectively a (ρ, ρ)-adapted (X, X)-matrix and a (σ, σ)-adapted (Y, Y)-matrix. Then the matrix BAC is also (ρ, σ)-adapted, and the persistence vector space $\theta(A)$ is isomorphic to $\theta(BAC)$.*

We will use this result to classify up to isomorphism all finitely presented persistence vector spaces. We begin by defining a persistence vector space $P(a, b)$ for every pair (a, b), where $a \in \mathbb{R}_+$, $b \in \mathbb{R}_+ \cup \{+\infty\}$, and $a < b$, with an obvious interpretation when $b = +\infty$. The space $P(a, b)$ is defined by $P(a, b)_r = \Bbbk$ for $r \in [a, b)$, $P(a, b) = \{0\}$ when $r \notin [a, b)$, and where the linear transformation $L(r, r') = id_{\Bbbk}$ whenever $r, r' \in [a, b)$. This definition can be interpreted in an obvious way when $b = +\infty$. We note that $P(a, b)$ is finitely presented. For, in the case where b is finite, let (X, ρ) and (Y, σ) denote $\mathbb{R}_+$-filtered sets (X, ρ) and (Y, σ), where the underlying sets consist of single elements x and y, with $\rho(x) = a$ and $\sigma(y) = b$. Then the 1×1 (X, Y)-matrix $[1]$ is (ρ, σ)-adapted, since $a \le b$, and it is clear that $P(a, b)$ is isomorphic to $\theta([1])$. When $b = +\infty$, $P(a, b)$ is isomorphic to the persistence vector space $V_{\Bbbk}(X, \rho)$ and can therefore be written as $\theta(0)$, where 0 denotes the zero linear transformation from the persistence vector space $\{0\}$.

Proposition 4.51   *Every finitely presented persistence vector space over $\Bbbk$ is isomorphic to a finite direct sum of the form*

$$P(a_1, b_1) \oplus P(a_2, b_2) \oplus \cdots \oplus P(a_n, b_n) \tag{4.1}$$

for some choices $a_i \in [0, +\infty)$, $b_i \in [0, +\infty]$, and $a_i < b_i$ for all i.

Proof   It is clear that a (ρ, σ)-adapted (X, Y)-matrix A which has the property that every row and column has at most one non-zero element, which is equal to 1, has the property that $\theta(A)$ is of the form described in the proposition. For, if we let $\{(x_1, y_1), (x_2, y_2), \dots, (x_n, y_n)\}$ be all the pairs (x_i, y_i) such that $a_{x_i, y_i} = 1$ then there is a decomposition

$$\theta(A) \cong \bigoplus_i P(\rho(x_i), \sigma(y_i)) \oplus \bigoplus_{x \in X - \{x_1, \dots, x_n\}} P(\rho(x), +\infty).$$

So, it suffices to construct matrices B and C which are respectively a (ρ, ρ)-adapted (X, X)-matrix and a (σ, σ)-adapted matrices (Y, Y)-matrix, such that BAC is has the property that every row and column has at most one non-zero element, which equals 1. To see that we can do this, we adapt the row and column operation approach to this setting. The (ρ, σ)-adapted row and column operations consist of all possible multiplications of a row or column by a non-zero element of k, all possible additions of a multiple of $r(x)$ to $r(x')$ when $\rho(x) \ge \rho(x')$, and all possible additions of a multiple of $c(y)$ to $c(y')$ when $\sigma(y) \ge \sigma(y')$. We claim that by performing (ρ, σ)-adapted row and column operations we can arrive at a matrix with at most one non-zero entry in each row and column. To see this, first find a y which maximizes $\sigma(y)$ over the set of all y with $c(y) \ne 0$. Next, find an x which maximizes $\rho(x)$ over the set of all x for which the entry $a_{xy} \ne 0$. Because of the way in which x is chosen, we are free to add multiples of $r(x)$ to all the other rows so as to cancel out $c(y)$ except in the xy-entry. Because of the way in which y is chosen, we can add multiples of $c(y)$ to cancel out $r(x)$ except in the xy-slot, without affecting $c(y)$. The result is a matrix in which the unique non-zero element in both $r(x)$ and $c(y)$ is a_{xy}. By multiplying $r(x)$ by $1/a_{xy}$, we can make the xy-entry in the transformed matrix equal to 1. By deleting $r(x)$ and $c(y)$, we obtain a $(X - \{x\}, Y - \{y\})$-matrix which is (ρ', σ')-adapted, where ρ' and σ' are the restrictions of ρ and σ to $X - \{x\}$ and $Y - \{y\}$, respectively. We can now apply the process inductively to this matrix. Each of the row and column operations required can be interpreted as operations on the row and columns of the original matrix and will have no effect on $r(x)$ or $c(y)$. The result is that by iterating this procedure we will eventually arrive at a matrix whose entries are equal to either zero or one, and it is clear that the transformed matrix has at most one non-zero entry in each row and column. The result now follows by Proposition 4.50.   □

We will also establish that any decomposition of the form (4.1) above for a given persistence vector space is essentially unique.

Proposition 4.52   *Suppose that $\{V_r\}$ is a finitely presented persistence vector space over k, and that we have two decompositions*

$$\{V_r\} \cong \bigoplus_{i \in I} P(a_i, b_i) \quad and \quad \{V_r\} \cong \bigoplus_{j \in J} P(c_j, d_j),$$

where I and J are finite sets. Then $\#(I) = \#(J)$, and the set of pairs (a_i, b_i) occurring in the decomposition, with multiplicities, is the identical to the set of pairs (c_j, d_j) occurring in the decomposition.

Proof   We let a_{min} and c_{min} denote the smallest values of a_i and c_j, respectively; a_{min} can be characterized intrinsically as $\min\{r | V_r \neq 0\}$, and it follows that $a_{min} = c_{min}$. Next, let b_{min} denote $\min\{b_i | a_i = a_{min}\}$, and make the corresponding definition for d_{min}; b_{min} is also defined intrinsically as $\min\{r' | N(L(r, r')) \neq 0\}$, where N denotes the null space, so $b_{min} = d_{min}$ as well. This means that $P(a_{min}, b_{min})$ appears in both decompositions. For each decomposition, we consider the sum of all occurrences of the summand $P(a_{min}, b_{min})$. They are both sub-persistence vector space of $\{V_r\}$, and can in fact be characterized intrinsically as the sub-persistence vector space $\{W_r\}$, where W_r is the null space of the linear transformation

$$im(L(a, r)) \xrightarrow{L(r,b) |_{im(L(a,r))}} V_b.$$

It now follows that the numbers of summands of the form $P(a_{min}, b_{min})$ in the two decompositions are the same, and further that they correspond isomorphically under the decompositions. Let I' denote the subset of I obtained by removing all i such that $a_i = a_{min}$ and $b_i = b_{min}$, and define J' correspondingly. We can now form the quotient of $\{V_r\}$ by $\{W_r\}$ and obtain the identifications

$$\{V_r\}/\{W_r\} \cong \bigoplus_{i \in I'} P(a_i, b_i) \quad and \quad \{V_r\}/\{W_r\} \cong \bigoplus_{j \in J'} P(c_j, d_j).$$

By induction on the number of summands in the decompositions, we obtain the result. □

The isomorphism classes of finitely presented persistence vector spaces are in one-to-one correspondence with finite subsets (with multiplicity) of the set $\{(a, b)|a \in [0, +\infty), b \in [0, +\infty]$, and $a < b\}$. Such sets can be represented visually in two distinct ways, one as families of intervals on the non-negative real lines and the other as a collection of points in the subset $\{(x, y)|x \geq 0$ and $y > x\}$ of the first quadrant in the (x, y)-plane. In the second case, one must place points with $b = +\infty$ above the whole diagram in a horizontal line indicating infinity. The first representation is called a *barcode*, and the second a *persistence diagram*; see Figure 4.9. We will use and refer to these representations interchangeably.

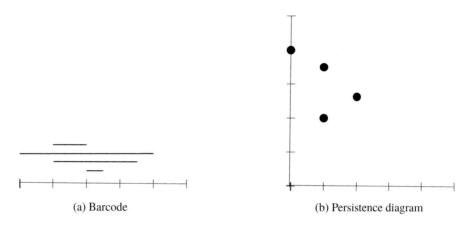

(a) Barcode   (b) Persistence diagram

Figure 4.9 Two methods for representing persistence vector spaces

4.5.1 Persistent Homology

We may at this point recall from Section 3.3.3 that from a simplicial complex Σ we may create a chain complex $C_*(\Sigma)$ as a free vector space on the simplices in Σ together with a boundary map ∂ defined on each simplex as the alternating sum of its top-dimensional faces. Homology groups then are defined as

$$H_*(\Sigma) = \operatorname{Ker} \partial / \operatorname{Im} \partial.$$

We can comfortably go through of these steps in the persistence setting: from a filtered simplicial complex Σ_*, we can create a persistent chain complex $C_*(\Sigma)$ by setting $C_*(\Sigma_*)_r = C_*(\Sigma_r)$: the persistent chain complex at parameter value r is the chain complex (in the classical sense) of the rth filtration step of Σ_*. Since, if $r < s$, the filtration provides inclusion maps $\Sigma_r \hookrightarrow \Sigma_s$, the functoriality of the chain complex construction gives us linear transformations of chain complexes $C_*(\Sigma_*)_r \hookrightarrow C_*(\Sigma_*)_s$. The boundary map is compatible with all this; $\partial : C_*(\Sigma_*) \to C_*(\Sigma_*)$ is a map of persistence vector spaces: the boundary of a simplex does not change just because we include it in a larger set of simplices. Hence, already the chain complex forms a persistence vector space. So do the kernel and the image, and quotients also respect the persistence structure. This means that $H_*(\Sigma_*)$ is also a persistence vector space, and as such has a decomposition that can be described by a barcode or a persistence diagram.

We now may associate with any finite metric space a persistence barcode or persistence diagram. What has now happened is that the Betti numbers have been replaced by barcodes. The way to reconcile these two notions is that, roughly speaking, persistence barcodes often consist of some "short" intervals and some "long" intervals. The short intervals are typically considered noisy, and the long ones are considered to correspond to larger-scale geometric features, which one would expect to correspond with the features of the space from which the metric space is sampled.

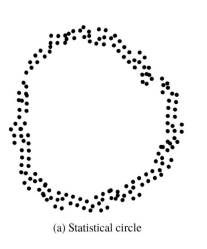

(a) Statistical circle

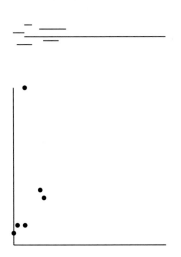

(b) Barcode (upper) and persistence diagram (lower) representations

The right-hand figure above shows the persistence barcode and persistence diagram for the one-dimensional homology associated with the sampled version of a circle (left-hand figure). The barcode reflects the fact that the first Betti number is equal to 1 by possessing a single long interval and multiple shorter ones.

Remark 4.53 Not all barcodes display this kind of dichotomous behavior between short and long. Such behavior reflects the fact that the metric space might not be representing a simple topological object of interest at a single scale, but rather a multiscale object in its own right. In addition, we will see in the next section that a number of different methods for generating barcodes can be devised; these, will reflect more subtle aspects of the shape of the data set.

Remark 4.54 It is clear from the descriptions given above that the complexity of persistent homology calculations is the same as that for Gaussian elimination, i.e. n^3 for an $n \times n$ matrix. What this means is that calculating directly will be extremely expensive. There are several approaches to mitigating this problem.

1. The α- and witness complexes, described in Sections 4.3.3 and 4.3.4, allow one to compute with much smaller complexes, based either on an embedding in a low-dimensional Euclidean space in the case of the α-complex or on a chosen set of landmarks in the case of the witness complex. Both are very practical options, and the same algorithms as those described above apply to that situation as well.
2. There are also methods for simplifying and drastically reducing the size of the Vietoris–Rips complex, described in Zomorodian (2010).
3. When we are given a space X with a finite covering by sets $\{U_\alpha\}_{\alpha \in A}$, there is a construction known as the *Mayer–Vietoris blowup* and a corresponding computational device known as the *Mayer–Vietoris spectral sequence* which permit the parallelization of the homology computation into calculations of much smaller complexity. We discuss some of this in Section 4.6. See Hatcher (2002) for a discussion of the case of a covering by two sets (the Mayer–Vietoris long exact sequence) and Segal (1968), Section 4, for the general case. This method proceeds by performing the individual calculations, and then provides a reconstruction step. This procedure was adapted to the persistent homology situation in Lipsky et al. (2011).

Remark 4.55   There are a number of software packages which compute homology and persistent homology. Among the older packages we find CHOMP (`http://chomp.rutgers.edu/`), Javaplex (`https://github.com/appliedtopology/javaplex/`), and Dionysus (`www.mrzv.org/software/dionysus/`). More recently we have the packages Phat (Bauer et al. 2017), Dipha (`https://github.com/DIPHA/dipha`), R-TDA (`https://CRAN.R-project.org/package=TDA`), and Gudhi (`https://gudhi.inria.fr/`). A good overview of the persistent homology software landscape can be found in Otter et al. (2017).

Remark 4.56   There are a number of theorems which produce theoretical guarantees for the computation of homology via various complexes. The so-called *nerve lemma* (Theorem 4.6), which follows directly from the construction in Segal (1968), Section 4, gives sufficient conditions for a much smaller construction based on a covering of the space to compute the homology accurately. In Niyogi et al. (2008), conditions are given which show that, with high confidence, a construction based on ϵ-balls around a finite sample from a submanifold of Euclidean space computes the homology of the submanifold accurately.

—————————————— For the advanced reader ——————————————

4.5.2   Direct Systems of Abelian Groups and Vector Spaces

In order to understand the structure of the collection of persistent homology groups, it is very convenient to package the groups up into a *direct system* of R-modules. By a direct system we simply mean a diagram indexed on $\mathbb{N}$: a collection of R-modules $\{M^i\}$ equipped with compatible maps of R-modules $f^i : M^i \rightarrow M^{i+1}$. The most important example in our present study is a direct system arising from a filtered complex.

Example 4.57   The homology of a filtered complex produces a direct system of R-modules, where the structure maps of the system are induced from the natural maps which take a cycle into the class containing it.

More generally, suppose we have a functor F from spaces (or simplicial complexes) to R-modules. We can regard a filtered complex as a direct system of simplicial complexes

$$K^0 \to K^1 \to K^2 \to \cdots \to K^n \to \cdots$$

Then applying F yields a direct system of R-modules

$$F(K_0) \to F(K_1) \to F(K_2) \to \cdots \to F(K^n) \to \cdots$$

by functoriality. For instance, the example above involves an application of the homology functor $H_n(-; R)$. Observe that when the coefficient ring R is $\mathbb{Z}$ we obtain a direct system of abelian groups, and when R is a field (typically $\mathbb{F}_p$ or $\mathbb{Q}$) we obtain a direct system of vector spaces.

Although often one considers the colimit associated with a direct system, in the present situation we will be primarily concerned with retaining all the information in the direct system. As such, we will be interested in the category of direct systems indexed on $\mathbb{N}$, where the morphisms are the natural transformations (i.e. a morphism $\{M_i\} \to \{N_i\}$ is specified by a collection of compatible maps $M_i \to N_i$). In the next subsection we will utilize the fact that the category of direct systems of vector spaces admits a concise classification (up to isomorphism).

4.5.3   Classification of Direct Systems of Vector Spaces

In order to have a meaningful classification, we need to impose some finiteness conditions. We say that a direct system $\{M_i\}$ of R-modules is of *finite type* if each constituent R-module M_i is itself finitely generated and if the system stabilizes in the sense that there exists an M such that the maps $M_j \to M_{j+1}$ are isomorphisms for all $j \geq M$. We will say that a filtered complex is of finite type under analogous conditions.

Example 4.58   A finite complex generates a filtered complex of chains which is of finite type and which produces a direct system of R-modules of finite type upon application of the homology functor.

Next, there is a natural functor Ψ from the category of direct systems of R-modules to the category of non-negatively graded $R[t]$-modules, defined on objects by taking direct systems $\{M_i\}$ to the direct sum

$$\bigoplus_{i=0}^{i=\infty} M_i,$$

where elements of R act componentwise and t acts by shifting via the structure maps of the direct system:

$$t(m_0, m_1, \ldots) = (0, f_0(m_0), f_1(m_1), \ldots).$$

The functor Ψ takes a natural transformation of direct systems to the componentwise map of direct sums; this is clearly compatible with the action of t and is therefore a graded $R[t]$-module map.

The Artin–Rees theorem implies that, when restricted to the full subcategory of direct systems of R-modules of finite type, the functor Ψ induces an equivalence of categories.

Theorem 4.59 *The functor Ψ induces an equivalence of categories between the category of direct systems of R-modules of finite type and the category of finitely generated non-negatively graded $R[t]$-modules.*

When R is a field, we can simplify this description further. In this case, $R[t]$ is a principal ideal domain (PID) and we know that the graded ideals are the homogeneous ideals (t^n) as n varies. By the structure theorem for PIDs, every finitely generated $R[t]$-module is isomorphic to a direct sum of $R[t]$-modules of the form

$$\left(\bigoplus_{i=1}^{n} \Sigma^{\alpha_i} R[t] \right) \oplus \left(\bigoplus_{j=1}^{m} \Sigma^{\gamma_j} R[t]/(t^{n_j}) \right),$$

where t^{n_j} divides $t^{n_{j+1}}$ and the Σ operation indicates a shift of grading.

4.5.4 Barcodes

There is a final reformulation of the classification of the persistent homological invariants associated with a filtered complex, which turns out to be an extremely useful way of visualizing the invariants. We continue to work in the setting where R is a field and so the structure theorem for the PID $R[t]$ applies. Let a $\mathcal{P}$-interval be a pair (i, j) where $0 \le i < j$ and $i, j \in \mathbb{Z} \cup \infty$. We associate a set of $\mathcal{P}$-intervals with a $R[t]$-module as follows: with each summand of the form $\Sigma^\alpha R[t]$, we associate the interval (α, ∞), and with each summand of the form $\Sigma^\beta R[t]/(t^n)$ we associate the interval $(\beta, \beta + n)$. This is clearly a bijection; given a set of $\mathcal{P}$-intervals, we associate $\Sigma^i R[t]$ with (i, ∞) and $\Sigma^i R[t]/(t^{j-i})$ with (i, j). This implies the following corollary.

Corollary 4.60 *There is a bijection between the finite sets of $\mathcal{P}$-intervals and the isomorphism classes of direct systems of R-modules of finite type.*

In the context of a filtered complex, this gives rise to a description of the persistent homology (as the filtration varies) over a field R as a *barcode*: each $\mathcal{P}$-interval (i, j) describes a cycle that appears at filtration i and specifies a homology class until it becomes a boundary at time j. This suggests a graphical representation of the persistent homology in terms of stacked intervals (i, j) or of points (i, j) in the upper half-plane $0 \le i \le j$.

4.5.5   Persistence and Noise in Topology

Algebraic topology is about the study of qualitative invariants of spaces, i.e. invariants which behave predictably in response to continuous deformation. One of the central goals of topological data analysis is to try to build qualitative invariants of point cloud data that are robust in the face of noise. For instance, one might hope that the process of associating a simplicial complex with a point cloud would take noisy perturbations of an underlying point cloud to homotopy equivalent complexes (and hence produce analogous topological invariants).

Unfortunately, this hope is wildly false. For any fixed ϵ, very small changes in the point cloud data can result in arbitrarily large deviations in the homological invariants of the point cloud. More precisely, given a fixed ϵ, two finite metric spaces which are close together in the Gromov–Hausdorff metric need not give rise to Rips or Čech complexes with similar Betti numbers. This is perhaps unsurprising, as we have already observed that in order for the homotopy type of the Rips and Čech complexes to be meaningful the scale parameter ϵ must be consistent with the underlying "feature scale" of the point cloud data.

Just as persistent homology provides a formal technique for coping with uncertainty about the appropriate feature scale for point cloud data, it turns out that persistent homology is an appropriate invariant to realize the vision of a robust topological invariant for a point cloud. Specifically, in this section, we discuss results of Memoli, Chazal, Cohen–Steiner, and collaborators, which we use to show that two finite metric spaces which are close together in the Gromov–Hausdorff metric have a bounded difference in the persistent homologies of their Rips complexes (in terms of a natural metric on barcodes). A large variety of such *stability theorems* have been proven, but we shall focus on this case.

We begin by discussing an alternative way of describing the persistent barcode associated the filtered complex. Suppose that we have a filtered complex K_α indexed on a collection of real numbers $\alpha_0 < \alpha_1 < \cdots < \alpha_n$ such that $K_{\alpha_0} = \emptyset$. As discussed, the k-homology groups of this filtered complex can be described as a barcode. An alternative description is in terms of a multiset of points in $\mathbb{R}^2$, where a point (x, y) encodes a homological feature which appears at time x and disappears at time y. For the purposes of stating the theorems, we introduce this in terms of sublevel sets of a continuous function, as follows.

Let X be a finite complex, and let $f : X \to \mathbb{R}$ be a function on X. Given a fixed integer k, define F_x to be the homology H_k of the inverse image $f^{-1}(-\infty, x]$. A useful example to keep in mind is the function f which assigns to the interior of each simplex the time at which it enters the filtration. We now define a *homological critical value* of f to be a real number a such that there exists k such that, for sufficiently small $\epsilon > 0$, the natural map $F_{a-\epsilon} \to F_{a+\epsilon}$ induced by the inclusion $f^{-1}(-\infty, a - \epsilon] \to f^{-1}(-\infty, a + \epsilon]$ is not an isomorphism.

Although we do not require that f be continuous, in order to work with its definition we need a finiteness condition called *tameness*: a function f is tame if there is a finite number of homological critical values and the ranks of the homology groups $H_k(F_x)$ are finite for all k and x. This implies that the associated persistence module has the finiteness conditions we discussed previously. A comprehensive study of different types of tameness conditions and their effect on persistent homologies can be found in Chazal et al. (2016).

Again fixing k, denote by f_x^y the natural map $F_x \to F_y$. Write F_x^y for the persistent homology group given as the image of F_x in F_y under f_x^y. We now describe a representation of persistence intervals in the form of a *persistence diagram*. Given a tame function $f\colon X \to \mathbb{R}$ with homological critical values $\{a_i\}$ and $\{b_i\}$ an interleaved sequence (satisfying $b_{i-1} < a_i < b_i$ for all i), set $b_{-1} = a_0 = -\infty$ and $b_{n+1} = a_{n+1} = \infty$. Then, for integer i, j such that $0 \le i < j \le n + 1$, define the multiplicity of the pair (a_i, a_j) to be the number

$$\mu_i^j = \dim F_{b_i-1}^{b_j} - \dim F_{b_i}^{b_j} + \dim F_{b_i}^{b_{j-1}} - \dim F_{b_i-1}^{b_{j-1}}.$$

The persistence diagram $D(f)$ of f is the multiset of points (a_i, a_j) counted with multiplicity μ_i^j along with all the points on the diagonal counted with infinite multiplicity. To interpret this in terms of our previous definitions, observe that when the filtration corresponds to a simplicial complex the multiplicity μ_i^j counts the classes which appear between F_{i-1} and F_i and vanish between F_{j-1} and F_j. That is, we are representing persistence intervals as points in the extended plane.

We now describe the metrics which arise in the statement of the main stability theorem. For two functions $f, g\colon X \to \mathbb{R}$, recall that the metric $||f - g||_\infty = \sup_x |f(x) - g(x)|$. Given multisets X and Y, we define the bottleneck distance as

$$d_B(X, Y) = \inf_\gamma \sup_x ||x - \gamma(x)||_\infty,$$

where $x \in X$ ranges over all points and γ ranges over all bijections from X to Y, each point with multiplicity k being treated as k distinct points. We can compute the bottleneck distance efficiently using bipartite matching algorithms.

Theorem 4.61 *Let X be a finite simplicial complex and let $f, g\colon X \to \mathbb{R}$ be continuous functions. Then the persistence diagrams associated with f and g satisfy $d_B(D(f), D(g)) \le ||f - g||_\infty$.*

This theorem, while very useful, relies on a fixed simplicial complex X. We now have the following extensions, which are most relevant for our setting.

Theorem 4.62 *Let (X, ∂_X) and (Y, ∂_Y) be finite metric spaces. Then for any k, we have the following bound:*

$$d_B(D_k R(X), D_k R(Y)) \le d_{GH}(X, Y),$$

where $d_{GH}(X, Y)$ is the Gromov–Hausdorff distance.

On the one hand, from a conceptual standpoint this theorem tells us that small perturbations in Gromov–Hausdorff space give rise to bounded changes in persistent homology. On the other hand, computing the Gromov–Hausdorff distance d_{GH} in practice is intractable, and therefore the bound of Theorem 4.62 gives us a way to estimate it.

We can extend this to the case of functions as follows. Denote by d^1_{GH} the extension of the Gromov–Hausdorff distance to metric spaces equipped with a continuous function $f: X \to \mathbb{R}$. Then we have the following extension of the theorem.

Theorem 4.63 *Let (X, ∂_X) and (Y, ∂_Y) be finite metric spaces, and $f: X \to \mathbb{R}$ and $g: Y \to \mathbb{R}$ continuous functions. Then for any k we have the following bound:*

$$d_B(D_k(X, f), D_k(Y, g)) \le d^1_{GH}((X, f), (Y, g)).$$

4.5.6   Persistent Cohomology

Just as in classical algebraic topology, we can take the filtered and persistence viewpoint and work with dualized vector spaces. Given a filtered simplicial complex Σ_* we can consider the cochains on Σ_*: the d-dimensional cochains $C^d(\Sigma_*)$ are functions from $C_*(\Sigma_*)$ to the constant persistent one-dimensional vector space $\Bbbk$. The maps $\Sigma_r \to \Sigma_s$ produce maps $C_*(\Sigma_*)_r \to C_*(\Sigma_*)_s$, which produce maps $C^*(\Sigma_*)_s \to C^*(\Sigma_*)_r$. The reason for this is that if $f: X \to Y$ and $g: Y \to Z$ are functions then $gf: X \to Z$ is a new function. The function f creates a way to transform any function $Y \to Z$ into a new function $X \to Z$, pulling the function g back along f. By taking f to be the linear map of chain complexes we see this order reversal emerge.

If we have picked bases for all the vector spaces, this dualization takes the shape of transposing the relevant matrices. From the boundary map ∂ we get the *coboundary map* $\delta = \partial^T$, where δf is the function that takes some chain z to the chain $\delta f(z) = f(\partial z)$.

With a coboundary in place, we can define cocycles $Z^* = \text{Ker } \delta$ and coboundaries $B^* = \text{Im } \delta$, and from these the cohomology groups $H^* = Z^*/B^*$. As we mentioned in Section 3.3.5, these cocycles and coboundaries end up obeying laws that are relatively easy to interpret. To illustrate, we will consider what happens in dimension 1:

A 1-cochain is a function from edges to $\Bbbk$.

A 1-cocycle is a function z such that $\delta z = 0$. Writing this out in more detail, this means that if $[abc]$ is a triangle in Σ_*, with edges $[ab], [ac]$, and $[bc]$, so that $\partial[abc] = [ab]-[ac]+[bc]$, then $\delta z([abc]) = z([ab]-[ac]+[bc]) = z([ab])-z([ac])+z([bc]) = 0$. Rewriting this we get $z([ab]) + z([bc]) = z([ac])$, telling us that cocycles are edge functions whose sum (corresponding to path integrals) are path independent between homotopic paths.

A 1-coboundary is a function z such that $z = \delta w$, in other words an edge function whose value is the difference in the values assigned to its endpoints. The wider analogy would be as a path integral of a potential field.

Persistent cohomology goes through all the required definitions and algebraic manipulations described in Section 4.5.1 and generates a persistent vector space with a barcode or persistence diagram description.

4.6   **Persistence and Localization of Features**

As shown in the previous section, one of the principal benefits of computing topological invariants of point clouds is their robustness to small local perturbations. In some sense, this reflects the difference between topological information and geometric information, in that topological invariants are more sensitive to global properties and less sensitive to specific local structure. However, sometimes it is still useful to be able to incorporate stronger notions of locality into the topological invariants we compute. For instance, non-trivial first Betti numbers tell us that a surface has holes. But we might be interested in gathering more information about the location of these holes.

In this section, we describe a method for using locality information (specified in the form of a cover) to recover information about the location of cycles in the homology of a point cloud. The method exploits persistence using a filtration that reflects the compatibility between different pieces of the cover to encode locality data; the persistence algorithm lets us align bases for the homology of the pieces of the filtration. Beyond its intrinsic interest, this example provides a good application of the idea of persistence (and persistent homology, specifically) to a different kind of domain from those discussed previously. We will see further examples of this sort of application later in the book.

Let X be an abstract simplicial complex, and suppose that it is covered by a family of subcomplexes $\mathcal{U} = \{U_i\}_{i=0}^n$. This means that every simplex of X is a simplex of at least one of the subcomplexes X_i. We formalize our localization problem as follows. We say that a homology class in $H_j(X)$ is $\mathcal{U}$-small if it is a sum of elements, each of which is in the image of the linear transformation $H_j(U_i) \to H_j(X)$ induced by the inclusion $U_i \hookrightarrow X$ for some i. We will show that this condition can be interpreted in terms of a certain persistence vector space.

We suppose that V_X is equipped with a total ordering. Let Δ^n denote the standard n-simplex, with vertex set $\{0, \ldots, n\}$ and with the standard total ordering. Using the two orderings on the vertex sets, we may form the product complex $\Delta^n \times X$, as in Section 3.2.7. For any subset $S \subseteq \{0, \ldots, n\}$, define $X[S]$ to be the subcomplex $\cap_{s \in S} X_s$. Also let $\Delta[S]$ denote the face of Δ^n associated with the subset S.

Definition 4.64   By the *Mayer–Vietoris blowup* of X attached to the covering $\mathcal{U}$, we mean the union of subcomplexes $\Delta[S] \times X[S]$. We denote it by $\mathfrak{M}(\mathcal{U})$. Projection on the X-factor produces a simplicial map

$$\pi: \mathfrak{M}(\mathcal{U}) \longrightarrow X.$$

Proposition 4.65   *The map π induces an isomorphism on the homology groups in all dimensions.*

Proof   This can be proved inductively on the skeleta of X, using the so-called long exact sequence of a pair. This is a computational technique which we have not included in this volume but which is introduced in Hatcher (2002). We omit the proof.   □

We define $\mathfrak{M}^{(k)}(\mathcal{U}) \subseteq \mathfrak{M}(\mathcal{U})$ to be the intersection $(\Delta^{(k)} \times X) \cap \mathfrak{M}(\mathcal{U})$ and observe that the filtration

$$\mathfrak{M}^{(0)}(\mathcal{U}) \subseteq \mathfrak{M}^{(1)}(\mathcal{U}) \subseteq \cdots \subseteq \mathfrak{M}^{(n)}(\mathcal{U}) = \mathfrak{M}(\mathcal{U}) \tag{4.2}$$

produces a persistence vector space for the k-dimensional homology group, for each k. We note that $\mathfrak{M}^{(0)}(\mathcal{U})$ is simply the disjoint union of the complexes U_i, and it is therefore easy to observe that

$$H_j(\mathfrak{M}^{(0)}(\mathcal{U})) \cong \oplus_i H_j(U_i).$$

It follows that an element $x \in H_k(X) \cong H_k(\mathfrak{M}(\mathcal{U}))$ is $\mathcal{U}$-small if and only if it is in the image of

$$H_k(\mathfrak{M}^{(0)}(\mathcal{U})) \to H_k(\mathfrak{M}^{(n)}(\mathcal{U})) \cong H_k(\mathfrak{M}(\mathcal{U})) \cong H_k(X). \tag{4.3}$$

The image of this map can be derived directly from the persistence vector space attached to the filtration (4.2) above as follows.

Because changes in the filtration occur only at integer values of the persistence parameter, the persistence barcode consists of bars $[i, j]$, where $0 \le i \le j \le n$ and i and j are integers. Tracing through the direct sum decomposition of the persistence vector space corresponding to the barcode, we determine a basis $\mathcal{B}$ for $H_k(X)$ which is in bijective correspondence with the collection of bars whose right-hand endpoint is n. We also find that a basis $\mathcal{B}_0$ for the image of the map (4.3) is given by the subset of $\mathcal{B}$ corresponding to intervals of the form $[0, n]$. The barcode also yields information about the homology elements in the various homology groups $H_k(U_i)$. A basis for the vector space $\bigoplus_i H_k(U_i)$ is in bijective correspondence with bars of the form $[0, i]$ for some i. The subset of that basis which corresponds to bars of the form $[0, n]$ consists of elements which map to the subset $\mathcal{B}_0$ of the basis $\mathcal{B}$. In summary, we have identified a basis for the subspace of elements which are $\mathcal{U}$-small, and further we have determined the origin of the classes.

We make a couple of remarks. First, sometimes we meet situations where, instead of simplicial complexes, we have more general topological spaces or spaces which are not equipped with a natural simplicial structure. In this case, there is a counterpart to Proposition 4.65 which considers more general spaces and coverings and which is introduced and discussed in Segal (1968). This counterpart to Proposition 4.65 permits the same kind of analysis using singular homology instead of simplicial homology. Second, the intermediate stages in the filtration (4.2) above are also of interest. One can consider the covering $\mathcal{U}[k] = \{U_\alpha\}_\alpha$, where α ranges over all sets $\{s_1, \ldots, s_k\}$ belonging to $\{0, \ldots, n\}$ that are of cardinality k, and where $U_\alpha = U_{s_1} \cup \cdots \cup U_{s_k}$. Our intuition suggests that homology classes in the kth filtration would be $\mathcal{U}[k]$-small. It turns out that that is not necessarily true, but it is true that homology classes that are $\mathcal{U}[k]$-small do lie in the kth filtration. We believe that more detailed analysis of the intermediate filtrations could give substantially more information.

4.7   Non-Homotopy-Invariant Shape Recognition

Homology is a very useful method for discriminating between shapes. However, there are many situations where it fails to address shape recognition problems that we might regard as qualitative.

Consider the spaces pictured in Figure 4.11. The space on the left is homotopy equivalent to a point, i.e. it is contractible and therefore has vanishing homology in all

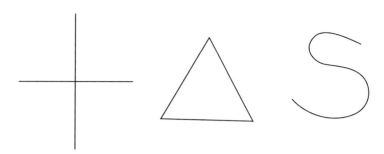

Figure 4.11 Shapes not adequately described by homology

positive dimensions. Nevertheless, we might very well be interested in the fact that it has four "ends", by which we mean the points at the tips of all four edges. The shape in the middle is homotopy equivalent to a circle, but we might be interested in the fact that it contains corners. Finally, the shape on the right is contractible, since it is homeomorphic to an interval. On the other hand, for many shape recognition problems, such as letter recognition, we might be interested in the fact that it is curved. Homology applied directly to these spaces will not yield a method for distinguishing them in the way we would like, since homology is homotopy invariant. We can address this problem in several ways.

1. We can construct filtrations on the space such that the sublevel sets are not contractible. The filtration will yield a persistence vector space, whose barcode will reflect the behavior of the sublevel sets.
2. We can construct auxiliary spaces whose homology reflects the qualitative behavior in which we are interested. An example we will discuss is the use of tangent information to construct the auxiliary space. It is possible for homology to detect the difference between the triangle on the one hand and the circle on the other.
3. We can construct an auxiliary space which is itself endowed with a filtration, so that persistent homology can detect the features in which we are interested.

Throughout this section, we will be dealing with actual topological spaces rather than point clouds. We will therefore use the convention that the term homology refers to singular homology. In later sections, we will introduce some discretization methods in particular situations.

4.7.1   Functional Persistence

Let X be a topological space, and let $f: X \to \mathbb{R}$ be a continuous function. We let $X(r) = f^{-1}((-\infty, r])$. Note that $X(r) \subseteq X(r')$ if $r \leq r'$. If we apply homology to these spaces, we obtain a persistence vector space

$$\{H_k(X(r))\}_r,$$

which has an attached barcode. This is called the *functional persistence barcode* attached to X and the function f.

A very useful function in a number of situations is a function that we will refer to as *eccentricity*. Suppose we are given a metric space which is bounded in the sense that there is an absolute upper bound for values of the distance function. Let δ denote the supremum of the values of the distance function. We will refer to it as the *diameter* of the space. For each point x, the set $\mathcal{D}(x) = \{d(x, y) \mid y \in X\}$ is bounded above, and we define $e(x)$ to be $\sup_x \mathcal{D}(x)$. This number is roughly a measure of the distance of the center of the space from a given point. In the persistence vector space we construct, we will want to introduce the topology of the points at the periphery first, and so it is useful to let

$$e^*(x) = \sup_x e(x) - e(x).$$

It is also useful to normalize the values of the function to lie between 0 and 1: we define $\overline{e}(x)$ by

$$\overline{e}(x) = \frac{e^*(x) - \inf_x e^*(x)}{\sup_x e^*(x) - \inf_x e^*(x)}.$$

This construction produces useful barcodes which allow us to distinguish between many different shapes which are in fact contractible.

Example 4.66   For the space on the left in Figure 4.11, we obtain a zero-dimensional barcode that looks as follows:

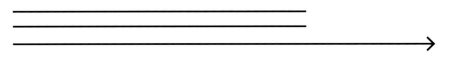

The four bars are tracking the four different connected components we obtain when we consider points which are sufficiently removed from the center. These components ultimately merge into a single point at the threshold value 1. One of the bars is infinite, and the other three have right-hand endpoint equal to 1. This barcode distinguishes the space in question from a point, and it also distinguishes it from an interval, which would produce a functional persistence barcode with two bars instead of four.

Example 4.67   Let $\mathfrak{H}$ denote the union of the three coordinate planes in $\mathbb{R}^3$ intersected by the unit ball. The intersection $\mathfrak{H}^0$ of the unit sphere with $\mathfrak{H}$ gives a space which is a union of three circles, each intersecting the other two in two points. One can perform this procedure easily using a homeomorphism from a simplicial complex to $\mathfrak{H}^0$. The simplicial complex in question has six vertices in bijective correspondence with the set $\{(\pm 1, 0, 0), (0, \pm 1, 0), (0, 0, \pm 1)\}$. There are 12 edges, corresponding to all pairs of vertices which are not antipodal pairs. There are a total of 15 pairs of vertices, of which three consist of antipodal pairs, so $15 - 3 = 12$ remain. There are no higher-dimensional simplices. We see that the space $\mathfrak{H}^0$ is connected, and it follows that the boundary operator has rank 5, and therefore its null space has dimension 7. This gives the first Betti number as 7. This means that we have seven bars with left-hand endpoint 0, and one can see that this Betti number persists until we reach 1, when all the loops vanish since the space has now become contractible. The one-dimensional barcode therefore consists of seven bars, each of which begins at 0 and ends at 7. Note that in this case the zero-dimensional barcode is trivial, i.e. it consists of a single infinite bar because all the spaces in question are connected.

Example 4.68 Consider the unit disc D^n in $\mathbb{R}^n$. We are free to construct the function $\bar{e}$ (see above), and evaluate the functional persistence for it. For $n > 1$, we find that the $(n-1)$-dimensional barcode for D^n contains a single bar of the form $[0, 1]$. The reason is that, for $0 < r < 1$, the space of points with $\bar{e}(x) \leq r$ is homotopy equivalent to the $(n-1)$-sphere. For $r = 0$, it is exactly the $(n-1)$-sphere, and otherwise we have that it is the n-dimensional analogue of an annulus and therefore homeomorphic to $S^{n-1} \times I$, where I denotes an interval.

Functional persistence can be used in other ways, when we are dealing with smooth objects but with varying notions of curvature.

Example 4.69 Suppose that we are considering the problem of distinguishing the letter C from the letter I. Both spaces are curves in the plane. We may compute the curvature $\kappa(x)$ at any point on the curve, and use it for functional persistence. Since the letter I has no curvature, its zero-dimensional persistent homology contains a single long infinite bar, beginning at the value 0. The letter C, however, has entirely positive curvature. Its zero-dimensional homology will also consist of a single bar, but beginning at a positive value, namely the minimum value of the curvature for the letter.

Example 4.70 Consider the ellipse given by the equation

$$\frac{x^2}{a^2} + \frac{y^2}{b^2} = 1,$$

with $a \geq b$. Calculus permits us to calculate the maximum and minimum values of curvature. We find that the maximum (respectively minimum) value of curvature is $\frac{a}{b^2}$ (respectively $\frac{b}{a^2}$), and occurs at the points $(\pm a, 0)$ (respectively $(0, \pm b)$). What this means is that the zero-dimensional barcode for this ellipse consists of two bars, with left-hand endpoints equal to $\frac{b}{a^2}$, one of the bars being infinite and the other ending at $\frac{a}{b^2}$. In the anomalous case of the circle, there is a single bar with left-hand endpoint at $\frac{1}{r}$, where r is the radius of the circle. This means that functional persistence effectively parametrizes the space of ellipses. Note that we regard two ellipses as equivalent if they one can be transformed to the other by a rigid transformation in the plane.

4.7.2 The Tangent Complex

In considering the triangle space in the center of Figure 4.11, we notice that one distinction between the triangle and the circle (which both have the same homology, since they are homeomorphic) is the fact that the triangle possesses corners while the circle does not. The goal of this section is to construct an auxiliary space whose homology does reflect the presence of the corners. The motivation for this construction comes from the construction of the tangent bundle to an embedded smooth k-dimensional submanifold M of $\mathbb{R}^n$ (see Milnor & Stasheff 1974). This produces two new manifolds, the *tangent bundle*, which has dimension $2k$ and is a so-called *vector bundle* over M, and the *unit tangent bundle*, which has dimension $2k - 1$ and which is an S^{k-1}-bundle over M. See Milnor & Stasheff (1974) for all these constructions. We will be dealing with subspaces of $\mathbb{R}^n$ that are not smooth submanifolds, but we generalize the construction as follows.

Definition 4.71   Let X be any subset of $\mathbb{R}^n$. We define $T^0(X) \subseteq X \times S^{n-1}$ as follows:

$$T^0(X) = \left\{ (x, \zeta) \left| \lim_{t \to 0} \frac{d(x + t\zeta, X)}{t} \right| = 0 \right\};$$

$T^0(X)$ is equipped with a projection $p \colon T^0(X) \to X$. We will refer to $T^0(X)$ as the *tangent complex* to X.

  This construction was used in Collins et al. (2004), Carlsson et al. (2005a), and Carlsson et al. (2005b), and is closely related to the notion of the *tangent cone* from geometric measure theory (Federer 1969). To understand the construction, we consider the case where X is the x-axis in $\mathbb{R}^2$. For each point on X, there are two possible choices of a vector ζ such that $(x, \zeta) \in T^0(X)$, namely $(1, 0)$ and $(-1, 0)$. One choice corresponds to a unit tangent direction in the positive x-direction and the other to one in the negative x-direction. Therefore $T^0(X)$ consists of two lines, one positive and the other negative. For the circle $X = S^1 \subseteq \mathbb{R}^2$, we find that $T^0(X)$ again breaks up as the disjoint union of two copies of S^1, one for the clockwise direction and the other for the counterclockwise direction. In both these cases, which involve smooth submanifolds of $\mathbb{R}^2$, $T^0(X)$ is actually an S^0-bundle over X, and we have $T^0(X) \cong X \times S^0$. Of course, S^0 consists of two points. When we have non-smooth submanifolds or non-manifold sets, $T^0(X)$ can become more interesting and useful. For example, let us consider the problem of distinguishing between the letter "U" and the letter "V". We can consider the letter U as a smooth curve in the plane, say as the set $X_U = \{(x, y) | x^2 + (y - 1)^2 = 1 \text{ and } 0 \le y \le 1\}$. On the other hand, the letter V can be represented by the set $X_V = \{(x, y) \mid |x| = |y| \text{ and } 0 \le y \le 1\}$. We see that X_U is a smooth submanifold of $\mathbb{R}^2$ and that $T^0(X_U)$ breaks into two distinct components, as in the cases we examined above. On the other hand, X_V breaks into four components, as shown in Figure 4.12. The point is that there are actually four distinct tangent directions at the point $(0, 0)$.

  As another example, the tangent complex to the triangle in Figure 4.11 consists of six disjoint intervals while the tangent complex to the circle consists of two disjoint circles. In both cases, we can detect the number of connected components using zero-dimensional homology.

Example 4.72   Consider the cone given by $\{(x, y, z) \mid x^2 + y^2 - z^2 = 0 \text{ and } z > 0\}$. For all points except $(0, 0, 0)$, the unit tangent space is a circle since the tangent space is a plane. However, at the origin, the tangent space is two-dimensional; in fact it is a two-dimensional torus $S^1 \times S^1$. The total tangent complex in this case is homeomorphic to the product space $[0, +\infty) \times S^1 \times S^1$. The space $[0, +\infty)$ can be thought of as an interval which is closed in one component of the boundary and which is open on the

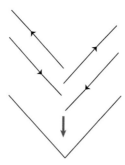

Figure 4.12 Tangent complex of the letter V. The red arrow indicates the projection from the tangent complex to the letter V itself

other component. If the cone point were smooth then the tangent complex would be $\mathbb{R}^2 \times S^1$. The homology definitely distinguishes between the two situations. For example, in the cone case the tangent complex has a non-vanishing two-dimensional homology while in the case of the smooth cone point, the complex is homotopy equivalent to the circle and therefore has no two-dimensional homology. This construction is an example of the *blowing up* construction favored in algebraic geometry (Hartshorne 1977).

4.7.3 Functional Persistence for Point Clouds

Applying persistent homology naively to many data sets will often produce trivial barcodes containing no long intervals. The reason for this is that data sets often have a central core to which everything is connected. In the data set in Figure 4.13, we see that there is a central core and apparently three "flares" emanating from it.

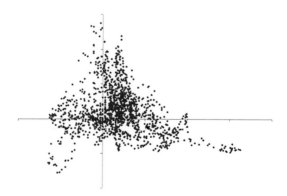

Figure 4.13 From http://macromarketmusings.blogspot.com/2007/09/tradeoff-between-output-price-stability.html

Roughly, this is a "T" or "Y" shape, and we would not expect to capture that aspect of its shape with homological methods since these spaces are contractible and therefore have vanishing homology. Similarly, if we were looking at a data set lying along a plane in a high-dimensional space, we would not expect to be able to detect that fact with homology, since a plane is also contractible. Our study of functional persistence in Section 4.7.1 suggests that we adapt that technique, which was used for actual topological spaces, to the point cloud situation. Here is a definition.

Definition 4.73 (Functional persistence for point clouds)   Let X be a finite metric space, and let $f : X \rightarrow \mathbb{R}$ be a non-negative real-valued function. Let us also select a positive real number ρ. Then by the *f-filtered simplicial complex with scale ρ* we will mean the increasing family of simplicial complexes

$$\{VR(f^{-1}([0, R]), \rho)\}_R.$$

Here R is the thresholding parameter, and the spaces $f^{-1}[0, R]$ grow with increasing R.

This construction will have persistence barcodes of its own, which reflect the topological properties of the sublevel sets of f. We will call this method of producing persistence diagrams *functional persistence for point clouds*.

Remark 4.74   Of course, in general it is not clear how to choose ρ. Often inspection of a data set can suggest the right scale, so that one can obtain useful information. Usually, though, it would be much better to construct two-dimensional profiles which encode both the function values and the scale at the same time. The kind of multidimensional persistence that would solve this problem is an object of current study (see Carlsson & Zomorodian 2009; Skryzalin & Carlsson 2017).

The eccentricity function $\bar{e}$ constructed in Section 4.7.1 can be used directly for finite metric spaces. It is part of a family of functions which can play a similar role, and one may find that for particular data sets some members of this family have more discriminatory power than others. The family consists of various functions that measure a notion referred to as the *data depth*, which are evaluated at a data point. For example, consider the family of functions e_p given by

$$e_p(x) = \sum_{x' \in X} d(x, x')^p.$$

We will want to introduce the larger values before the smaller values of a function. For this reason we will replace e_p by the function $\bar{e}_p$ given by

$$\bar{e}_p(x) = \frac{e_p^{\max} - e_p(x)}{e_p^{\max} - e_p^{\min}},$$

where $e_p^{\max}$ and $e_p^{\min}$ are the maximum and minimum values taken by the function e_p on X. The function $\bar{e}_p$ takes values on $[0, 1]$ and attains both both 0 and 1. The value $p = +\infty$ gives the function $\bar{e}$ introduced in Section 4.7.1.

Example 4.75   The data set in the following diagram is shaded according to the values of $\bar{e}_p$, with light shading corresponding to low values of $\bar{e}_p$ and dark shading to high values.

For a small value of R, the sublevel set of $\bar{e}_p$ would look as follows:

and so the zero-dimensional barcode would be of the form

Example 4.76   This data set is also shaded by values of $\bar{e}_p$:

A sublevel set of $\bar{e}_p$ for a small value of R would be

Since this point cloud is roughly circular, the one-dimensional barcode would look like

In both cases, the behavior of the set of points "far from the center" measures an interesting aspect of the shape of a data set; in the one case we have a set of clusters and in the other a circle.

Another class of functions which are useful for the study of data are density estimators, i.e. values of any proxy function for density, such as the codensity function for finite metric spaces introduced in Section 2.8. Typically we will want to include denser points earlier than sparser points, and so if we were given a density function ρ on our data set we might use the function

$$\xi(x) = \frac{\rho_{\max} - \rho(x)}{\rho_{\max} - \rho_{\min}}$$

as our persistence parameter, where $\rho_{\max}$ and $\rho_{\min}$ denote $\sup_x \rho(x)$ and $\sup_x \rho(x)$, respectively, assuming the values $\rho_{\max}$ and $\rho_{\min}$ exist. For finite metric spaces, they clearly do. If we are dealing finite metric spaces, the kth codensity proxy $\delta_k(x)$ from Section 2.8 is inversely related to density, and we would use the function

$$\eta(x) = \frac{\delta_k(x) - \delta_k^{\min}}{\delta_k^{\max} - \delta_k^{\min}}$$

for the persistence parameter. These density-related functions give useful information about the density function. As an example, consider the density function on the real line pictured in the image in Figure 4.14(a).

The zero-dimensional persistence barcode for the function ξ defined above would contain two bars, both beginning at the value zero, one being infinite and the other ending roughly at the value $\frac{1}{2}$. The barcode tracks the components of the superlevel sets of the density function, and it is clear that at the value zero there are two connected components of this set. The two connected components merge roughly halfway through the range of ξ. The distribution shown in Figure 4.14(b) will similarly produce four bars, one infinite and beginning at zero and the other three beginning at some small positive values and ending

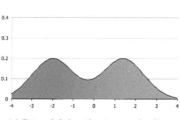

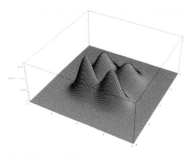

(a) Bimodal distribution on the line   (b) Multimodal distribution in the plane

Figure 4.14 Images from `https://en.wikipedia.org/wiki/Multimodal_distribution`. (a) published under CC BY-SA 3.0.

at some larger positive values. What this shows is that zero-dimensional persistence barcodes can be used to detect *modes*, i.e. local maxima of the density function.

Density functions can be more complicated, where there might be a whole space of local maxima. For example, suppose that we have a probability distribution in the plane whose probability density function is proportional to

$$\frac{1}{1 + d(\vec{x}, S^1)^2},$$

where S^1 is the unit circle. In this case, the entire circle consists of maxima for the probability density function; the zero-dimensional persistence barcode does not see any interesting properties of the density function but the one-dimensional barcode does.

4.8   Zig-Zag Persistence

Persistence objects (sets, vector spaces, etc.) are defined as a collection of objects parametrized by the non-negative real line together with morphisms from objects with parameter value r to objects with parameter value r', whenever $r \leq r'$. If we restrict the persistence object to the lattice of non-negative integers, we can view a persistence object as a diagram

$$X_0 \to X_1 \to X_2 \to \cdots \to X_n \to \cdots$$

of objects X_n and morphisms $X_n \to X_{n+1}$ for all $n \geq 0$. We can informally state that a persistence object restricted to the integers is equivalent to a diagram of objects having the shape

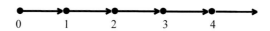

where the nodes are objects (sets, vector spaces, simplicial complexes, ...) and the arrows indicate a morphism from one object to another. As such, it is an infinite *quiver diagram* (Derksen & Weyman 2005). Any directed graph Γ is called a *quiver*, and a representation of Γ over a field $\Bbbk$ is an assignment of a $\Bbbk$-vector space V_v to each vertex of Γ and of a $\Bbbk$-linear transformation $L_e : V_v \to V_w$ to every edge e from v to w. We will also refer to a representation of Γ as a *diagram of shape* Γ. One could also consider diagrams with the shape given below:

$$0 \qquad 1 \qquad 2 \qquad 3 \qquad 4$$

This corresponds precisely to a family of objects parametrized by the non-negative integers, with a morphism $X_i \to X_{i+1}$ when i is even and a morphism $X_{i+1} \to X_i$ when i is odd. We will call such a diagram of $\Bbbk$-vector spaces, where $\Bbbk$ is a field, a *zig-zag persistence vector space*. Zig-zag vector spaces arise in a number of ways.

Example 4.77   Consider a very large finite metric space X, so large that we do not expect to be able to compute its persistent homology using a Vietoris–Rips complex. We might instead try to form many smaller samples $\{S_i\}$ from X, and attempt to understand how consistent these procedures are. We note that given such a family of samples, we may fix a threshold parameter R and construct $\text{VR}(S_i, R)$ for all i. To attempt to assess the consistency of the computations which arise out of these complexes, we can form the unions $S_i \cup S_{i+1}$ and note that we have inclusion maps $S_i \hookrightarrow S_i \cup S_{i+1}$ and $S_{i+1} \hookrightarrow S_i \cup S_{i+1}$. These maps induce maps on the Vietoris–Rips complexes, and, by applying homology to the Vietoris–Rips complexes, we obtain $\Bbbk$-vector spaces $V_i = H_j(\text{VR}(S_i, R))$ and $V_{i,i+1} = H_j(\text{VR}(S_i \cup S_{i+1}, R))$. The inclusion maps now mean that we obtain a diagram of the form

$$V_0 \to V_{0,1} \leftarrow V_1 \to V_{1,2} \leftarrow V_2 \to V_{2,3} \leftarrow V_3 \to \cdots .$$

On an intuitive level, consistency between the calculations is measured by the existence of classes $x_i \in V_i$ and $x_{i+1} \in V_{i+1}$ such that the images of x_i and x_{i+1} in $V_{i,i+1}$ are in the same non-zero class, and more generally, for sequences of classes $\{x_i\}$, with $x_i \in V_i$, such that for all i the images of x_i and x_{i+1} in $V_{i,i+1}$ are equal to the same non-zero element.

Example 4.78   Given a simplicial complex X and a simplicial map from X to the non-negative real line, triangulated so as to allow the vertex set to be the non-negative integers and the edges to be the closed intervals $[n, n+1]$, we can form the subcomplexes $f^{-1}([n, n+1])$, as well as $f^{-1}(n)$. We have a diagram

$$f^{-1}(0) \to f^{-1}([0, 1]) \leftarrow f^{-1}(1) \to f^{-1}([1, 2])$$
$$\leftarrow f^{-1}(2) \to f^{-1}([2, 3]) \leftarrow f^{-1}(3) \to \cdots .$$

A scheme which could compute the homology only of the complexes based at the nodes of this diagram and which could extract the homology of the entire complex from these calculations would permit the parallelization of homology computations into smaller pieces. This would be very desirable, of course.

Example 4.79   In the discussion of the witness complexes, we selected a set of landmarks $\mathcal{L}$ from a metric space X, and computed certain complexes $W(X, \mathcal{L}, \epsilon)$ for which there were several variants. In general, it is difficult to assess how accurately the persistent homology of X is captured by a witness complex. One piece of evidence for accuracy would be a method which assesses the consistency of different choices of landmarks. In order to do this, we constructed (in Section 4.3.4) a bivariate version of the C^V-construction which takes as input a pair of landmark sets $(\mathcal{L}_1, \mathcal{L}_2)$ and is denoted by $W(X, (\mathcal{L}_1, \mathcal{L}_2), \epsilon)$, and which maps to each of the complexes $W(X, \mathcal{L}_1, \epsilon)$ and $W(X, \mathcal{L}_2, \epsilon)$ in a natural way. Given a collection of landmark sets $\mathcal{L}_i$, one obtains a diagram of witness complexes

$$W(X, \mathcal{L}_1, \epsilon) \leftarrow W(X, (\mathcal{L}_1, \mathcal{L}_2), \epsilon) \rightarrow W(X, \mathcal{L}_2, \epsilon)$$
$$\leftarrow W(X, (\mathcal{L}_2, \mathcal{L}_3), \epsilon) \rightarrow W(X, \mathcal{L}_3, \epsilon) \leftarrow .$$

After applying homology, one can again ask for consistent families as in Example 4.77.

It turns out that there is a classification theorem for zig-zag persistence vector spaces over a field $\Bbbk$ which is analogous to that for ordinary persistence.

Definition 4.80   A zig-zag persistence $\Bbbk$-vector space V, comprising a family of k-vector spaces as in Definition 4.44, is called *cyclic* if there are integers $m \leq n$ such that $V_i = \Bbbk$ for $m \leq i \leq n$, $V_i = \{0\}$ if $i < m$ or $i > n$, and the homomorphisms $V_i \rightarrow V_{i+1}$ or $V_{i+1} \rightarrow V_i$ are all equal to the identity homomorphism on $\Bbbk$ whenever $m \leq i \leq i + 1 \leq n$.

An example is the following; the superscript *id* indicates the identity mapping.

$$\{0\} \rightarrow \Bbbk \xleftarrow{id} \Bbbk \xrightarrow{id} \Bbbk \leftarrow \{0\} \rightarrow \{0\} \leftarrow \cdots .$$

Here *id* is the identity map from k to itself. Each cyclic persistence vector space is indecomposable, in that it cannot be expressed as a direct sum of diagrams. Note that cyclic zig-zag persistence vector spaces are parametrized by intervals $[m, n]$ with integral endpoints. We write $V[m, n]$ for the cyclic $\Bbbk$-vector space which is non-zero exactly for the integers in $[m, n]$. We have the following analogue of Theorem 4.51.

Theorem 4.81   *We say that a zig-zag persistence k-vector space V is of finite type if (a) each vector space V_i is finite-dimensional and (b) $V_i = \{0\}$ for sufficiently large i. Then, for any zig-zag persistence $\Bbbk$-vector space V of finite type, there is an isomorphism*

$$V \cong \bigoplus_i V[m_i, n_i] \tag{4.4}$$

for some choices of pairs of integers (m_i, n_i). Furthermore, the decomposition (4.4) above is unique up to a reordering of the summands.

The content of this theorem is that there is a barcode description of the isomorphism classes of zig-zag persistence vector spaces which is just like that for ordinary persistence vectors spaces except that the intervals in the barcodes are constrained to be integers. This theorem was first proved by P. Gabriel (1972) and was discussed from the point of view of computational issues in Carlsson & de Silva (2010). This result can now be applied to Examples 4.77 and 4.79 above. The presence of long bars in the barcode decomposition says that there are elements which are consistent across all the vector spaces in the interval determined by the bar. For Example 4.78 above, the zig-zag construction can be used to give a very efficient and parallelizable method for computing the homology of the entire complex. These ideas were discussed in Carlsson & de Silva (2010) and Carlsson et al. (2009).

4.9   Multidimensional Persistence

One of the problems which frequently comes up in persistent homological calculations is that the presence of a few outliers can mask the structure. The example below is suggestive.

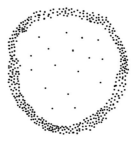

Note that the main structure is a sampling from a loop, but that there are a number of outliers in the interior of the loop. The homology of the complexes built from this data set will not strongly reflect the loop, because the short connections among the outliers and between the outliers and the points on the actual loop will quickly fill in a disc. In order to remedy this, we would have to find a principled way to remove the outliers in the interior. A measure of density would have this effect, because the outliers by many measures of density would have a much lower density than the points on the loop itself. So, if one selected, say, the 90% densest points by a density measure, one would effectively be choosing the loop itself. However, this choice of threshold is arbitrary, so just as we chose to maintain a profile of homology groups over all threshold values for the scale parameter, we might attempt to create such a profile for values of density.

Another problem with persistent homology has already been made apparent in Section 4.7.3, where we constructed the increasing family of simplicial complexes $\{VR(f^{-1}([0, R]), \rho\}_R$ for a non-negative real-valued function f on our data set and for a *fixed* choice of our scale parameter ρ. These complexes allow us to study topological features of the data sets which are not captured by the direct application of persistent homology, but using them requires us to make a choice of the scale parameter. There is no universal or natural choice of ρ, so it would be very desirable to track both R and ρ simultaneously.

These considerations motivate the following definition. We let $\mathbb{R}_+ = [0, +\infty)$ and define a partial order on $\mathbb{R}_+^n$ by declaring that $(r_1, \ldots, r_n) \leq (r'_1, \ldots, r'_n)$ if and only if $r_i \leq r'_1$ for all $1 \leq i \leq n$.

Definition 4.82 By an *n-persistence vector space* over a field $\Bbbk$, we mean a family of $\Bbbk$-vector spaces $\{V_{\vec{r}}\}_{\vec{r} \in \mathbb{R}^n}$ together with linear transformations $L(\vec{r}, \vec{r}') \colon V_{\vec{r}} \to V_{\vec{r}'}$ whenever $\vec{r} \leq \vec{r}'$, so that

$$L(\vec{r}', \vec{r}'')L(\vec{r}, \vec{r}') = L(\vec{r}, \vec{r}'')$$

whenever $\vec{r} \leq \vec{r}' \leq \vec{r}''$. There are obvious notions of linear transformations and isomorphisms of n-persistence vector spaces.

We might hope that there is a compact representation of the isomorphism classes of n-persistence vector spaces that is analogous to the barcode or persistence diagram representations available in 1-persistence, i.e. the ordinary persistence we have already discussed. It turns out, though, that this is not possible, for the following reason. As was observed in Zomorodian & Carlsson (2005), ordinary persistence vector spaces have much in common with the classification of graded modules over the graded ring $\Bbbk[t]$. Indeed, if we restrict the domain of the scale parameter r in ordinary persistence to the integer lattice $\mathbb{Z}_+ \subseteq \mathbb{R}_+$ then the classification of such restricted persistence vector spaces is identical to the classification of graded $\Bbbk[t]$-modules. Similarly, it was shown in Carlsson & Zomorodian (2009) that the isomorphism classes of n-persistence vector spaces with parameter set restricted to $\mathbb{Z}_+^n$ is identical to the classification of n-graded modules over the n-graded ring $\Bbbk[t_1, \ldots, t_n]$. It is well understood in algebraic geometry that the classification of modules over polynomial rings in more than one variable is fundamentally different from the one-variable case. In particular, for graded one-variable polynomial rings, the parametrization of the set of isomorphism classes is independent of the underlying field $\Bbbk$, whereas this does not hold for multigraded polynomial rings in more than one variable. In fact, the classification for more than one variable usually involves *spaces* of structures rather than discrete sets. What these problems suggest is that we should drop the idea of dealing with a complete classification of the isomorphism classes of n-persistence vector spaces, and instead develop useful invariants which we expect will measure useful and interesting properties of n-persistence vector spaces.

One approach to finding invariants is via the functions to be defined in Section 5.2. It turns out that some of them can be interpreted in ways which do not depend on obtaining an explicit barcode representation. For any finitely presented persistence vector space $\{V_r\}_r$, we can define two functions attached to V, $\Delta_V \colon \mathbb{R}_+ \to \mathbb{R}_+$ and $\rho_V \colon \mathbb{R}_+^2 \to \mathbb{R}_+$. We set

$$\Delta_V(r) = \dim(V_r)$$

and

$$\rho_V(r, r') = \operatorname{rank}(L(r, r')).$$

Example 4.83 We consider the function τ_{10} from Section 5.2, which is given by $\sum_i (y_i - x_i)$. It is easy to check that the equation

$$\tau_{10}(x_1, y_1, \ldots, x_n, y_n) = \int_{\mathbb{R}_+} \Delta_V(r)\,dr$$

holds for finite barcodes.

Example 4.84   One can also show that

$$\frac{1}{2}(\tau_{10}^2 - 2\tau_{20}) = \frac{1}{2}\sum_i (y_i - x_i)^2 = \int \int_{\mathbb{R}_+^2} \rho_V(r, r')dr\,dr'.$$

Example 4.85   We also have

$$\tau_{11} = \int_{\mathbb{R}_+} r\Delta_V(r)dr.$$

None of the integrals involves the explicit barcode decomposition; they depend only on information concerning the dimensions of the spaces V_r and the ranks of the linear transformations $L(r, r')$. The value of this is that Δ_V and ρ_V have direct counterparts in n-persistence situations, and integrating them (this time over $\mathbb{R}_+^n$ and $\mathbb{R}_+^{2n}$, respectively) yields analogues of at least these invariants in the multidimensional situation. The study of invariants of multidimensional persistence vector spaces is the subject of ongoing research. Our approach here was studied in detail in Skryzalin & Carlsson (2017).

5 Structures on Spaces of Barcodes

5.1 Metrics on Barcode Spaces

We have now associated with any finite metric space a family of persistence barcodes, or persistence diagrams. Since the Vietoris–Rips complex is much easier to compute, we will default to Vietoris–Rips persistence barcodes unless we explicitly state otherwise. One important property to understand is the degree to which the barcode changes when we have small (in a suitable sense) changes in the data. In order even to formulate the answer to such a question, we will need to define what is meant by small changes in the barcode. In order to do this, we will define the *bottleneck distance* between barcodes. First, for any pair of intervals $I = [x_1, y_1]$ and $J = [x_2, y_2]$, we will define $\Delta(I, J)$ to be the ℓ^∞-distance between the two, regarded as ordered pairs in $\mathbb{R}^2$, i.e. $\max(|x_2 - x_1|, |y_2 - y_1|)$. For a given interval $I = [x, y]$, we also define $\lambda(I)$ to be $\frac{y-x}{2}$. Here $\lambda(I)$ is the ℓ^∞-distance to the closest interval of the form $[z, z]$ to I. Given two families $\mathbb{I} = \{I_\alpha\}_{\alpha \in A}$ and $\mathbb{J} = \{J_\beta\}_{\beta \in B}$ of intervals, for finite sets A and B and any bijection θ from a subset $A' \subseteq A$ to $B' \subseteq B$, we will define the *penalty* of θ, $P(\theta)$, to be

$$P(\theta) = \max\left(\max_{a \in A'}(\Delta(I_a, J_{\theta(a)})), \max_{a \in A - A'}(\lambda(I_a)), \max_{b \in B - B'}(\lambda(J_b))\right).$$

We then define the bottleneck distance $d_\infty(\mathbb{I}, \mathbb{J})$ to be

$$\min_\theta P(\theta),$$

where the minimum is taken over all possible bijections from subsets of A to subsets of B. This is easily verified to be a distance function on the set of barcodes.

Remark 5.1 The bottleneck distance d_∞ is actually the $p = \infty$ version of a family of metrics d_p, called the Wasserstein metrics. The metric d_p is defined via a penalty function P_p given by

$$P_p(\theta) = \sum_{\alpha \in A'} \Delta(I_\alpha, J_{\theta(\alpha)})^p + \sum_{\alpha \in A - A'} \lambda(I_\alpha)^p + \sum_{b \in B - B'} \lambda(J_b)^p$$

and we set $d_p(\mathbb{I}, \mathbb{J}) = (\min_\theta P_p(\theta))^{1/p}$. We will also refer to d_p as the p-Wasserstein distance.

We now have a notion of what it means for barcodes to be close. There is also a notion of what it means for two compact metric spaces to be close, given by the *Gromov–Hausdorff* distance, first defined in Burago et al. (2001). It is defined as follows. Let Z

be any metric space, and let X and Y be two compact subsets of Z. Then the *Hausdorff distance* between X and Y, $d^{\mathrm{H}}(X, Y)$, is defined to be the quantity

$$\max \left\{ \max_{x \in X} \min_{y \in Y} d_Z(x, y), \max_{y \in Y} \min_{x \in X} d_Z(x, y) \right\}.$$

Given any two compact metric spaces X and Y, we consider the family $\mathcal{I}(X, Y)$ of all simultaneous isometric embeddings of X and Y. An element of $\mathcal{I}(X, Y)$ is a triple (Z, i_X, i_Y), where Z is a metric space and i_X and i_Y are isometric embeddings of X and Y, respectively, into Z. The Gromov–Hausdorff distance of X and Y is now defined to be the infimum over $\mathcal{I}(X, Y)$ of $d^{\mathrm{H}}(\mathrm{Im}(i_X), \mathrm{Im}(i_Y))$. It is known to give a metric on the collection of all compact metric spaces. It is also known to be computationally very intractable. In Chazal et al. (2009), the following is proved.

Theorem 5.2 *Fix a non-negative integer i, let $\mathcal{F}$ denote the metric space of all finite metric spaces, and let $\mathcal{B}$ denote the set of all persistence barcodes. Let $\beta_k \colon \mathcal{F} \to \mathcal{B}$ be the function which assigns to each finite metric space its k-dimensional homology barcode. Then β_k is distance non-increasing.*

This result is interesting not only because it gives some guarantees on how certain changes in the data affect the result, but also because it provides an easily computed lower bound on the Gromov–Hausdorff distance in a great deal of generality.

There are also stability results for functional persistence or, rather, an analogue of functional persistence. In Definition 4.73, we defined functional persistence using a function on the points of a metric space, or equivalently on the vertices of its Vietoris–Rips complex. For any topological space X and continuous real-valued function $f \colon X \to \mathbb{R}$, one can also define associated persistence vector spaces $\{H_i(f^{-1}((-\infty, r])\}_r$ which are very close analogues of the simplicial complex construction described above. In fact, a real-valued function on the vertex set of a simplicial complex can be extended in a natural way to a continuous function on the geometric realization of the simplicial complex, using a weighted sum of values of the function on the vertices on the basis of the *barycentric coordinates* of the point. The idea will be that small changes in the function should yield only small changes in the associated persistence barcode. The results we will describe appeared in Cohen-Steiner et al. (2007) and Cohen-Steiner et al. (2010).

To state the results, we will need some definitions. Let X be a topological space, and let $f \colon X \to \mathbb{R}$ be a real-valued function on X. For every non-negative integer k, $a \in \mathbb{R}$, and $\epsilon \in (0, +\infty)$, we have the induced map

$$j = j_{k,a,\epsilon} \colon H_k(f^{-1}(-\infty, a - \epsilon]) \longrightarrow H_k(f^{-1}(-\infty, a + \epsilon]).$$

We say a is a *homological critical value* of f if there is a k such that $j_{k,a,\epsilon}$ *fails* to be an isomorphism for all sufficiently small ϵ. Further, we say the function f is *tame* if it has a finite number of homological critical values and the homology groups $H_k(f^{-1}(-\infty, a])$ are finite-dimensional for all $k \in \mathbb{N}$ and $a \in \mathbb{R}$. This condition holds, for example, in the case of Morse functions on closed manifolds (see Milnor 1963) and for piecewise linear functions on finite simplicial complexes, so the result we will state is quite generally applicable. The following theorem is proved in Cohen-Steiner et al. (2007).

Theorem 5.3 *Let X be any space which is homeomorphic to a simplicial complex, and suppose that $f, g: X \to \mathbb{R}$ are continuous tame functions. Then the persistence vector spaces $\{H_k(f^{-1}((-\infty, r])\}_r$ and $\{H_k(g^{-1}((-\infty, r])\}_r$ are finitely presented and therefore admit barcode descriptions for each $k \in \mathbb{N}$, which we denote by $\beta_k f$ and $\beta_k g$. Moreover, for any k we have that*

$$d_\infty(\beta_k f, \beta_k g) \leq \|f - g\|_\infty.$$

Furthermore, a stability result for the Wasserstein distances d_p with p finite, in the presence of a Lipschitz property for the functions, was proved in Cohen-Steiner et al. (2010).

5.2 Coordinatizing Barcode Space and Feature Generation

One of the core lessons we may learn from algebraic geometry is the fundamental role played by *coordinate maps* on geometric structures: for a space $\mathbb{X}$, a coordinate map is a continuous map $\mathbb{X} \to \mathbb{Y}$ to some simpler and better understood space $\mathbb{Y}$.

Coordinate maps are important because they give us a way to look at the shape of $\mathbb{X}$ through lenses we understand well, and to handle the information describing $\mathbb{X}$ through tools developed for $\mathbb{Y}$. We can view coordinate maps as a way of describing *feature generation* procedures: with a well-chosen $\mathbb{Y}$, the features generated by coordinate maps form a digestible input to more classical statistical or machine learning techniques. We will lay out several such ways of coordinatization for feature generation in the following sections, and revisit everything in Section 6.6 where the methods we describe here are leveraged for concrete applications.

5.2.1 Symmetric Polynomials

Another way to describe infinite sets is via the theory of *algebraic varieties*, i.e. as the set of solutions to a set of equations, over the real numbers, the complex numbers, or some other field. When this is possible, it gives a very compact description of a large or infinite set. The method also produces a ring of functions on the set, by restricting the polynomial functions to it. We will discuss a coordinatized model of the set of all barcodes, and the resulting ring of functions.

Let us consider first the set $\mathbb{B}_n$ of all barcodes containing exactly n intervals or "bars". Each of the intervals is determined by two coordinates, the left-hand endpoint x and the right-hand endpoint y. If we have n intervals, we have $2n$ coordinates $\{x_1, y_1, \ldots, x_n, y_n\}$. The trouble is that the barcode space does not take into account the ordering of the intervals, so it is not possible to assign a value to the ith coordinate itself. To understand how to get around this problem, we discuss a familiar situation from invariant theory.

We will consider $\mathbb{R}^n$ and let Σ_n be the group of permutations of the set $\{1, \ldots, n\}$. We also consider the ring of polynomial functions $A_n = \mathbb{R}[x_1, \ldots, x_n]$ on $\mathbb{R}^n$. We will describe how to coordinatize the set of orbits (as defined in Section 3.2.5) $\mathbb{R}^n / \Sigma_n$, i.e. the "set of unordered n-tuples of real numbers", or equivalently the collection of multisets of size n. If f is a polynomial function on $\mathbb{R}^n$, it may be treated as a function on the set of orbits if it has the property that $f(\sigma\vec{v}) = f(\vec{v})$ for all $\sigma \in \Sigma_n$ and $\vec{v} \in \mathbb{R}^n$. The group Σ_n acts on the ring A_n, and an element $f \in A_n$ is a function on the orbit set $\mathbb{R}^n/\Sigma_n$ if and only if it is fixed under all permutations $\sigma \in \Sigma_n$. The set of all fixed functions (denoted by $A_n^{\Sigma_n}$) is a ring in its own right, and it turns out that it has a very simple description.

Proposition 5.4 *The ring $A_n^{\Sigma_n}$ is isomorphic to the ring of polynomials*

$$\mathbb{R}[\sigma_1, \sigma_2, \ldots, \sigma_n],$$

where σ_i denotes the ith elementary symmetric function given by

$$\sum_{s_1, s_2, \ldots, s_i} x_{s_1} x_{s_2} \cdots x_{s_i},$$

and where the sum is over all i-tuples of distinct elements of $\{1, \ldots, n\}$.

We now have coordinates which describe the set $\mathbb{R}/\Sigma_n$. The analogous construction for the set $\mathbb{B}_n$ would be to form the subring of $\mathbb{R}[x_1, y_1, \ldots, x_n, y_n]$ that is fixed under the action of Σ_n which permutes the x_i among themselves, and the y_i among themselves and so to obtain a coordinatization this way. This does give a ring of functions; it is not a pure polynomial ring, however, but has *relations*, or *syzygies*. A full discussion of this situation was given in Dalbec (1999). To give an idea of what this means, we first observe that the elementary symmetric functions $\sigma_i(\vec{x})$ and $\sigma_i(\vec{y})$ are definitely invariant, and generate a full polynomial subring of the ring of functions. Let us restrict to the case $n = 2$. Then there is another function $\xi = (x_1 y_1 + x_2 y_2)$ which cannot be expressed in terms of the elementary symmetric functions applied to $\vec{x}$ and $\vec{y}$. It turns out that there is now an algebraic relation

$$\xi^2 - \sigma_1(\vec{x})\sigma_1(\vec{y})\xi + \sigma_1(\vec{x})^2\sigma_2(\vec{y}) + \sigma_2(\vec{x})\sigma_1(\vec{y})^2 - 4\sigma_2(\vec{x})\sigma_2(\vec{y}) = 0,$$

which, after consideration, shows that there is algebraic coordinatization not by the four-dimensional affine space but rather by a subset of a higher-dimensional space cut out by one or more algebraic equations. Nevertheless, one is able to express $\mathbb{B}_n$ as the points of an algebraic variety in this way.

What we would really like to do is to coordinatize the collection of all the sets $\mathbb{B}_n$ as a variety, in an appropriate sense. One could consider the disjoint union $\coprod_n \mathbb{B}_n$, but we would rather disregard intervals of length zero in a systematic way, since they correspond to features which are born and die at the same time and hence do not have any actual persistence. This suggests that we define a set $\mathbb{B}_\infty$ as follows. We let $\sim$ denote the equivalence relation on $\coprod_n \mathbb{B}_n$ generated by the equivalences

$$\{[x_1, y_1], [x_2, y_2], \ldots, [x_n, y_n]\} \sim \{[x_1, y_1], [x_2, y_2], \ldots, [x_{n-1}, y_{n-1}]\}$$

whenever $x_n = y_n$. We then define the infinite barcode set $\mathbb{B}_\infty$ to be the quotient

$$\coprod_n \mathbb{B}_n / \sim .$$

The question we now pose is whether this infinite set can also be coordinatized, again in a suitable sense. To see how this can be done, we consider two simpler examples of coordinatization.

Example 5.5 We let $\mathfrak{A}_n$ denote the set $\mathbb{R}^n / \Sigma_n$ and $\mathfrak{A}_\infty$ denote

$$\coprod_n \mathfrak{A}_n / \sim,$$

where $\sim$ is the equivalence relation given by declaring that $(x_1, \ldots, x_n) \sim (y_1, \ldots, y_m)$ if and only there are sets $S \in \{1, \ldots, n\}$ and $T \subseteq \{1, \ldots, m\}$ satisfying the following properties.

1. $x_s = 0$ and $y_t = 0$ for all $s \in S$ and $t \in T$.
2. We have $n - \#(S) = m - \#(T)$, and we will call this common number k.
3. The unordered k-tuples obtained by deleting the elements corresponding to S and T from $(x_1, \ldots, x_n)$ and $(y_1, \ldots, y_m)$, respectively, are identical.

There are natural maps from $\mathfrak{A}_n$ to $\mathfrak{A}_\infty$, and the maps are injective on points. We therefore have an increasing system

$$\mathfrak{A}_1 \hookrightarrow \mathfrak{A}_2 \hookrightarrow \mathfrak{A}_3 \hookrightarrow \cdots,$$

which corresponds to a system of ring homomorphisms

$$\mathbb{R}[\sigma_1] \leftarrow \mathbb{R}[\sigma_1, \sigma_2] \leftarrow \mathbb{R}[\sigma_1, \sigma_2, \sigma_3] \leftarrow \cdots,$$

where the homomorphism $\mathbb{R}[\sigma_1, \ldots, \sigma_{n+1}] \to \mathbb{R}[\sigma_1, \ldots, \sigma_n]$ is defined by $\sigma_i \to \sigma_i$ for $1 \leq i \leq n$, and $\sigma_{n+1} \to 0$.

Associated with such a system is its *limit*, which is itself a ring. We will not go into detail on this, but refer to MacLane (1998) for background material on limits and their dual construction, a colimit. In this case, the construction produces a ring of functions which can be described as follows. Let M denote the set of all monomials in the infinite set $\{\sigma_1, \sigma_2, \ldots, \sigma_n, \ldots\}$, and let $M_n \subseteq M$ denote the subset of monomials which involve only $\{\sigma_1, \ldots, \sigma_n\}$. Then the inverse limit ring is identified with the set of all infinite sums $\sum_{\mu \in M} r_\mu \mu$ such that the sums $\sum_{\mu \in M_n} r_\mu \mu$ are all finite. So, for instance, the infinite sum $\sum_n \sigma_n$ is an element of this ring. Elements in this ring certainly define functions on $\mathfrak{A}_\infty$, because the functions σ_N vanish on $\mathfrak{A}_n$ whenever $n \leq N$. We will regard the functions σ_i as coordinates on $\mathfrak{A}_\infty$, with the understanding that, for any point $x \in \mathfrak{A}_\infty$, the vector $(\sigma_1(x), \sigma_2(x), \ldots)$ has the property that only finitely many coordinates are non-zero.

In the example above, taking the quotient by the equivalence relation $\sim$ did not create complicated rings, as each of the rings $\mathbb{R}[\sigma_1, \sigma_2, \ldots, \sigma_n]$ is a pure polynomial ring in n variables. In general, though, taking quotients by equivalence relations can create rings which are not pure polynomial, and in some cases are not even finitely generated as algebras.

Example 5.6 Consider the plane $X = \mathbb{R}^2$, and consider the equivalence relation $\sim$ defined by declaring that $(x, 0) \sim (x', 0)$ for all x, x'. This relation "collapses" the entire x-axis to a point, while leaving the rest of the set unchanged. The question is whether $\mathbb{R}^2 / \sim$ can be described as an algebraic variety. As in the case where the ring of invariants of a group action gives a variety structure on an orbit set, in the present case we will ask which polynomial functions f on $\mathbb{R}$ have the property that $f(x) = f(x')$ whenever $x \sim x'$. This means that we are asking for polynomials f in two variables such that $f(x, 0) = f(x', 0)$ for all x, x'. A quick calculation shows that this ring of functions consists of all polynomials in x and y such that the linear term in x is zero. So, a basis for it is the set of monomials $\{x^i y^j \mid i > 0 \implies j > 0\}$. This is a ring which is easy to understand, but it is not a finitely generated algebra. It is generated by the elements $\theta_i = x^i y$ together with the element y, and they satisfy the relations $\theta_i^2 = y\theta_{2i}$. We can then obtain a description of $\mathbb{R}^2/\sim$ as the set of points in $\mathbb{R}^\infty = \{(y, \theta_1, \theta_2, \ldots) \mid \theta_i^2 = y\theta_{2i}$ for all $i > 0\}$. Note that in this case infinitely many coordinates will typically be non-zero for a given (x, y).

The image in the map $\mathbb{B}_n \rightarrow \mathbb{B}_\infty$, which we denote by $\mathbb{B}'_n$, can be described as the set obtained from $\mathbb{B}_n$ by taking the quotient by an equivalence relation $\sim_n$ defined as follows. Given two multisets of intervals (each with n intervals) $S = \{[x_1, y_1], [x_2, y_2], \ldots, [x_n, y_n]\}$ and $S' = \{[x'_1, y'_1], [x'_2, y'_2], \ldots, [x'_n, y'_n]\}$, we say that $S \sim_n S'$ if there are subsets $I, I' \subset \{1, \ldots, n\}$ such that $x_i = y_i$ for all $i \in I$ and $x'_{i'} = y'_{i'}$ for all $i' \in I'$, and such that the multisets $S - \{[x_i, x_i] \mid i \in I\}$ and $T - \{[x'_{i'}, x'_{i'}] \mid i' \in I'\}$ are identical multisets of intervals. This means that $\mathbb{B}'_n$ can be expressed as the quotient of $\mathbb{B}_n$ by an equivalence relation similar to that described in Example 5.6. It corresponds to a ring $A(\mathbb{B}'_n)$, which is in general not finitely generated. However, we consider the increasing sequence of sets

$$\mathbb{B}'_1 \hookrightarrow \mathbb{B}'_2 \hookrightarrow \mathbb{B}'_3 \hookrightarrow \cdots$$

and a corresponding system of ring homomorphisms

$$A(\mathbb{B}'_1) \longleftarrow A(\mathbb{B}'_2) \longleftarrow A(\mathbb{B}'_3) \longleftarrow \cdots,$$

as in Example 5.5. This system also has an limit, which we will denote by $A(\mathbb{B}_\infty)$ and which can be described as follows. Let $\mathcal{N}$ denote the set of all monomials in a set of variables $\{\tau_{ij} \mid 1 \leq i, 0 \leq j\}$, and let $\mathcal{N}_k \subseteq \mathcal{N}$ denote the subset of monomials in $\{\tau_{ij} \mid 1 \leq i \leq k, j \geq 0\}$. Then the ring $A(\mathbb{B}_\infty)$ is identified with the set of all infinite sums $\sum_{\nu \in \mathcal{N}} r_\nu \nu$ which have the property that all the sums $\sum_{\nu \in \mathcal{N}_k} r_\nu \nu$ are finite sums. The variables τ_{ij} correspond to functions on $\mathbb{B}_\infty$ which can be described as follows. It will suffice to described τ_{ij} on $\mathbb{B}_n$, i.e. on an unordered n-tuple of intervals. To do this, we first define a function τ'_{ij} on the set of *ordered* n-tuples on intervals by

$$\tau'_{ij}([x_1, y_1], \ldots, [x_n, y_n]) = (y_1 - x_1) \cdots (y_i - x_i) \left(\frac{y_1 + x_1}{2}\right)^j.$$

To obtain τ_{ij} we simply symmetrize by writing

$$\tau_{ij} = \sum_{\sigma \in \Sigma_n} \tau'_{ij} \circ \sigma.$$

So, for example, τ_{10} applied to an unordered n-tuple of interval is the sum of the lengths of the intervals, τ_{20} is the second elementary symmetric function in the lengths, and τ_{1j} is the sum over all the intervals of the product of the length of the interval with the jth power of its midpoint.

These functions, while useful, are not continuous when the space of barcodes is equipped with the bottleneck distance. Kališnik (2019) has studied a *tropical* analogue. Recall that tropical algebraic geometry (see Maclagan & Sturmfels 2015) studies semirings of functions where the sum operation (respectively the multiplication operation) is replaced by max or min (respectively addition). It was shown that the study of symmetric tropical polynomials follows the same pattern as ordinary symmetric polynomials (Carlsson & Kališnik Verovšek 2016), but that the ring of tropical functions on spaces of barcodes is too small, in the sense that it does not separate barcodes. This means that there are pairs of distinct barcodes whose values under every symmetric tropical polynomial are identical. Kališnik (2019) showed that one can resolve this problem by introducing tropical rational functions, and she found that they do indeed separate points.

5.2.2 Persistence Landscapes

The barcode or persistence diagram descriptions of persistent homology offer themselves as natural representations of the topological structures of a point cloud. However, they are in themselves not particularly amenable to statistical analysis. There have been efforts underway since at least 2017 to produce statistical approaches to persistence diagrams: through density estimation images (Adams et al. 2017), through cumulative distribution functions along interval midpoints (Biscio & Møller 2016), and through central limit theorems (Duy et al. 2016). One of the very first approaches, however, was introduced in Bubenik (2015) and generates a function $\mathbb{N} \times \mathbb{R} \to \mathbb{R} \cup \{-\infty, \infty\}$ from any persistence diagram. Statistical analyses then can be made to work on the pointwise means of these new functions.

Definition 5.7 (Bubenik 2015) The persistence landscape of a persistence diagram $\{(b_i, d_i)\}$, which need not be finite or even countably infinite, is the function

$$\lambda(k, t) = \sup \{m \geq 0 \mid (t - m, t + m) \text{ is a subinterval of at least } k \text{ intervals}\}.$$

These functions are everywhere non-negative, weakly decreasing in k, and 1-Lipschitz: the change in the landscape is bounded by the change in the underlying data.

Bubenik (2015) went on to prove a central limit theorem. To do this, he first fixed an L_p-norm $\| - \|_p$ to work with, and observed that, since we can view point clouds as random variables, their corresponding persistence landscapes are also random variables. So, for a sample $\mathbb{X}_1, \ldots, \mathbb{X}_n$ of point clouds, producing landscapes $\lambda_1, \ldots, \lambda_n$, we can consider $\bar{\lambda}_n = n^{-1} \sum_j \lambda_j$.

Theorem 5.8 (Law of large numbers) $\bar{\lambda}_n \to \mathbb{E}\lambda$ almost surely if and only if $\mathbb{E}\|\lambda\|_p$ is finite.

Theorem 5.9 (Central limit) If $p \geq 2$ and $\mathbb{E}\|\lambda\|_p$ as well as $\mathbb{E}(\|\lambda\|_p^2)$ are finite, then $\sqrt{n}\left(\bar{\lambda}_n - \mathbb{E}\lambda\right)$ converges weakly to a Gaussian random variable with the same covariance structure as λ.

Picking a function $f \colon \mathbb{N} \times \mathbb{R} \to \mathbb{R}$ with finite L_q-norm, for q such that $1/p + 1/q = 1$, we can use f to assign weights to the $(\mathbb{N} \times \mathbb{R})$-plane and produce a single-value statistic

$$Y = \int_{\mathbb{N}\times\mathbb{R}} f\lambda = \|f\lambda\|_1.$$

Then from the central limit theorem we obtain

$$\sqrt{n}\left(\bar{Y} - \mathbb{E}Y\right) \to \mathcal{N}(0, \text{var}\, Y).$$

From this normal distribution we can derive confidence intervals and hypothesis testing in the usual way; any introductory textbook in statistics will give an outline of the process.

5.2.3   Persistence Images

Persistence images are another vectorization of persistence diagrams. They were introduced by Adams et al. (2017), who proceeded by treating a persistence diagram as a collection of point masses, and then smoothing the corresponding distribution using some auxiliary data. The auxiliary data consists of two items.

1. A function $\phi\colon \mathbb{R}^2 \times \mathbb{R}^2 \to \mathbb{R}$ such that the function $\phi(u, -)$ is a probability distribution on $\mathbb{R}^2$ for all u, and such that the mean of the function $\phi(u, -)$ is u.
2. A continuous and piecewise differentiable non-negative weighting function $f\colon \mathbb{R}^2 \longrightarrow \mathbb{R}$ which takes the value zero along the x-axis.

A common choice for ϕ is to set $\phi(u, -)$ equal to the probability density function for the standard spherically symmetric Gaussian distribution with mean equal to u and with variance equal to a fixed σ.

We will find it useful to recoordinatize the persistence diagram via the coordinate change $(x, y) \to (x, y - x)$. The reason for this transformation is that the persistence diagram consists of a set of points lying in the set

$$\{(x, y) \mid x, y \geq 0 \text{ and } y \geq x\} = (\xi, \eta),$$

which is transformed bijectively to the first quadrant by the transformation. We have a transformed collection of points making up the persistence diagram, which we denote by $\{(\xi_1, \eta_1), \ldots, (\xi_n, \eta_n)\}$ and refer to as $\mathcal{B}$. From this set, we define the function $\rho_{\mathcal{B}}\colon \mathbb{R}^2 \to \mathbb{R}$ by the formula

$$\rho_{\mathcal{B}}(\vec{z}) = \sum_{i=1}^{n} f(\xi_i, \eta_i)\phi((\xi_i, \eta_i), \vec{z}).$$

This is a function on the upper half-plane which can be regarded as an image by encoding the numerical values using a gray scale, hence the name. We often want finite-dimensional vector representations rather than objects in a function space. One way to achieve this, described in Adams et al. (2017), is to first restrict the function to a finite rectangle in the plane, next divide that rectangle into a grid of rectangular subregions, and then create a vector with a coordinate for each of the subregions; the coordinates are the integrals of $\rho_{\mathcal{B}}$ over the subregions. Once this is done, we obtain a map PI: $\mathbb{B}_\infty \to \mathbb{R}^N$. The following continuity property is proved in Adams et al. (2017).

Theorem 5.10   *The map* PI *is continuous when the metric on* $\mathbb{B}_\infty$ *is the 1-Wasserstein distance.*

For the advanced reader

5.3   Distributions on $\mathbb{B}_\infty$

We have seen that one can use persistence barcodes to obtain invariants of finite metric spaces which mimic homological invariants for ordinary topological spaces. For finite metric spaces obtained by sampling, one can hope to perform inference on the barcode invariant. For example, if one sees a barcode with a bar which one perceives as "long", can one determine whether such a long bar could have occurred by chance. More generally, how can one use barcodes to reject a null hypothesis that the sample was obtained from a fixed distribution or family of distributions? In order to perform such an inference, one will need to develop a theory of probability measures on $\mathbb{B}_\infty$, and in Mileyko et al. (2011) some important first steps in this direction were taken. We summarize their work.

The set $\mathbb{B}_\infty$ becomes a topological space on being equipped with the quotient topology (see Section 3.2.5) under the map

$$\coprod_n \mathbb{B}_n \to \mathbb{B}_\infty,$$

where each $\mathbb{B}_n$ is topologized using the quotient topology for $\mathbb{R}^{2n} \to \mathbb{R}^{2n}/\Sigma_n$, and where $\coprod_n \mathbb{B}_n$ is topologized by declaring that a set in $\coprod_n \mathbb{B}_n$ is open if and only if its intersection with $\mathbb{B}_n$ is open for each n. In Section 5.1 metrics d_p, with the possiblity of $p = \infty$, were introduced on $\mathbb{B}_\infty$. The goal of the work in Mileyko et al. (2011) was to study the possibility of defining expectations and variances, in the Fréchet sense (see Fréchet, 1944, 1948), on the metric spaces $(\mathbb{B}_\infty, d_p)$. One obstacle to realizing this goal is the fact that the metric space $(\mathbb{B}_\infty, d_p)$ is not complete. In order to deal with this problem, Mileyko et al. (2011) constructed completions $\hat{\mathbb{B}}_p$ of $(\mathbb{B}, d_p)$. We will first enlarge the set $\mathbb{B}_\infty$ to include countable multisets of intervals $\{I_a\}_{a \in A}$, where A is a countable set. For a fixed p, the underlying set of $\mathbb{B}_p$ will be the set of all $\{I_a\}_{a \in A}$ for which

$$\sum_{a \in A} \lambda(I_a) < +\infty. \tag{5.1}$$

It is now clear by inspection that the distance d_p extends naturally to a metric $\hat{d}_p$ on $\mathbb{B}_p$ via the same formula as in Section 5.1, where the sum may be infinite but is always convergent owing to the condition (5.1). We now have the following.

Theorem 5.11 (Mileyko et al. 2011)   *The metric space $(\hat{\mathbb{B}}_p, \hat{d}_p)$ is complete and separable.*

The goal of Mileyko et al. (2011) was to construct means and variances on $\hat{\mathbb{B}}_p$. In general metric spaces X there is no notion of a single point-valued mean; rather, the mean will actually be a subset of X. Here are the definitions of Fréchet means and variances, paraphrased from Mileyko et al. (2011).

Definition 5.12 Let X be a metric space and $\mathcal{B}(X)$ its Borel σ-algebra. Let $\mathcal{P}$ be a probability measure on $(X, \mathcal{B}(X))$, and suppose that $\mathcal{P}$ has a finite second moment, i.e. that $\int_X d(x, x')^2 d\mathcal{P}(x') < \infty$ for all $x \in X$. Then by the Fréchet variance of $\mathcal{P}$ we mean

$$\mathrm{Var}_\mathcal{P} = \inf_{x \in X} \left[F_\mathcal{P}(x) = \int_X d(x, x')^2 d\mathcal{P}(x') \right]$$

and we will call the set

$$\mathbb{E}_\mathcal{P} = \{x \mid F_\mathcal{P}(x) = \mathrm{Var}_\mathcal{P}\}$$

the Fréchet expectation or Fréchet mean of $\mathcal{P}$.

As stated, $\mathbb{E}_\mathcal{P}$ always exists as a set but it may be empty or contain more than one point. There are results on the non-emptiness and the uniqueness of Fréchet means for manifolds (Karcher 1977; Kendall 1990). What was proved in Mileyko et al. (2011) is an existence result for the case $X = \hat{\mathbb{B}}_p$.

Theorem 5.13 *Let $\mathcal{P}$ be a probability measure on $(\hat{\mathbb{B}}_p, \hat{d}_p)$, and suppose that $\mathcal{P}$ has finite second moment and compact support. Then $\mathbb{E}_\mathcal{P} \neq \emptyset$.*

In Turner et al. (2014) this existence theorem is made algorithmic, but for a somewhat different choice of metric and metric space. By the L^2-*Wasserstein metric* on $\mathbb{B}_\infty$, we mean the analogue of d_2 for the choice of penalty function

$$\lambda_{L^2}(I, J) = (x_1 - x_2)^2 + (y_1 - y_2)^2.$$

The resulting metric space is simpler to work with computationally, and all the existence results from Mileyko et al. (2011) were also shown to hold in Turner et al. (2014).

Another approach concerns the notion of distances between measures on metric spaces. It was developed in Blumberg et al. (2014) and Chazal et al. (2011). One useful choice is the so-called *Lévy–Prohorov metric*.

Definition 5.14 Given a metric space (X, d_X), we define the Lévy–Prohorov metric π_X on the set $\mathcal{P}(X)$ of probability measures on the measurable space X, with its σ-algebra of Borel sets $\mathcal{B}(X)$, by

$$\pi_X(\mu, \nu) = \inf\{\epsilon > 0 \mid \mu(A) \leq \nu(A^\epsilon) + \epsilon \text{ and } \nu(A) \leq \mu(A^\epsilon) + \epsilon \text{ for all } A \in \mathcal{B}(X)\},$$

where A^ϵ denotes the ϵ-neighborhood of A, i.e. the union of all balls of radius ϵ about points of A. This metric is defined for all metric spaces, and is known to induce the topology of weak convergence of measures on $\mathcal{P}(X)$ when X is a "Polish space", i.e. a separable complete metric space.

By a *metric measure space* (X, d_X, μ_X), we mean a metric space (X, d_X) together with a probability measure μ_X on the σ-algebra of Borel sets associated with the metric d_X. In Greven et al. (2009), an analogue to the Gromov–Hausdorff metric is defined on the collection of compact metric measure spaces: given two metric measure spaces $X = (X, d_X, \mu_X)$ and $\mathcal{Y} = (Y, d_Y, \mu_Y)$, the *Gromov–Prohorov distance* between X and $\mathcal{Y}$ is defined in Greven et al. (2009) to be

$$d_{\mathrm{GPr}}(X, \mathcal{Y}) = \inf_{(\varphi_X, \varphi_Y, Z)} \pi_{(Z, d_Z)}((\varphi_X)_* \mu_X, (\varphi_Y)_* \mu_Y),$$

where, as in the definition of the Gromov–Hausdorff metric, $(\varphi_X, \varphi_Y, Z)$ varies over all pairs of isometric embeddings $\varphi_X \colon X \to Z$ and $\varphi_Y \colon Y \to Z$ of X and Y into a compact metric space Z.

The distributions on $\mathbb{B}_\infty$ of interest in Blumberg et al. (2014) are those which are obtained by sampling n points in X, according to the probability measure μ_X, and computing a persistent homology in a fixed dimension k to obtain persistence barcodes. The idea is that each one is a kind of finite approximation to the metric measure space, and the result shows that the differences between these measures are controlled by the Gromov–Prohorov distance between the actual metric spaces. More precisely, given a metric measure space (X, d_X, μ_X), these authors constructed probability measures $\Phi^n_k = \Phi^n_k(X, d_X, \mu_X)$ on the completed barcode spaces $\hat{\mathbb{B}}_\infty$ described above, via the formula

$$\Phi^n_k(X, d_X, \mu_X) = (\beta_k)_*(\mu^n_X),$$

where β_k denotes the k-dimensional barcode and $(-)_*$ denotes the push-forward measure on $\hat{\mathbb{B}}_p$. This formula makes sense because (a) β_k is a continuous map of metric spaces by Theorem 5.2 and (b) both spaces are given the structure of measurable spaces via Borel σ-algebras. The main theorem of Blumberg et al. (2014) is now the following.

Theorem 5.15 (Blumberg et al. 2014, Theorem 5.2)   *Let (X, d_X, μ_X) and (Y, d_Y, μ_Y) be compact metric measure spaces. Then we have the following inequality relating the Prohorov and Gromov–Prohorov metrics:*

$$d_{\mathrm{Pr}}(\Phi^n_k(X, d_X, \mu_X), \Phi^n_k(Y, d_Y, \mu_Y)) \le n d_{\mathrm{GPr}}((X, d_X, \mu_X), (Y, d_Y, \mu_Y)).$$

This estimate permits one to prove the following convergence result.

Corollary 5.16 (Blumberg et al. 2014, Corollary 5.5)   *Let $S_1 \subset S_2 \subset \cdots \subset \cdots$ be a sequence of randomly drawn samples from (X, d_X, μ_X). We regard S_i as a metric measure space using the subspace metric and the empirical measure. Then $\Phi^n_k(S_i)$ converges in probability to $\Phi^n_k(X, d_X, \mu_X)$.*

Remark 5.17   One could formulate similar results in other contexts. For example, a very interesting object of study would be to study the distributions obtained by choosing instead various landmark sets in a witness complex. This would then give an assessment of how well the space is represented by witness complexes of a fixed size.

Part III

Practice

Practice

6 Case Studies

To showcase the techniques discussed earlier on real datasets, we will introduce a number of published research projects and explain how they have used persistent (co)homology or the Mapper algorithm in active research.

6.1 Mumford Natural Image Data

Consider an image produced by a black and white digital camera. It consists of a rectangular array of pixels, together with a numerical value called the *gray-scale value* for each pixel, which encodes the shade of gray as intermediate between black and white. Although it is convenient to think of the gray scale as taking continuous values, it is actually encoded using a fixed number of bits, commonly 8 or 16. It is possible to regard each image as a vector in a space of dimension P, where P is the number of pixels. Creating a collection of images can then be viewed as a representation of the visual world, or some part of it. On the other hand, neuroscientists have obtained a degree of understanding of the workings of the visual pathway in the brain. One of the better understood portions of the visual pathway is the so-called *primary visual cortex* or *V1*. This is made up of a large array of neurons, which have distinct functions. An important study of the function of individual neurons was initiated by Hubel & Wiesel (1964). One finding has been that neurons often have an orientation-specific sensitivity, so that they might respond to a straight edge between a light and a dark region at a particular angle. In addition, there are neurons which specifically respond to a dark line on top of a light background, but not to a boundary between two regions.

It is an interesting question to contemplate how any structure one might find in this space of images would correlate with the neuron-by-neuron understanding that one has about V1 from the work of Hubel and Wiesel. One might hypothesize, for example, that frequently occurring patterns in small regions of the visual field should have associated with them "dedicated" neurons, which fire in the presence of that particular pattern. As a step in this direction, two Dutch neuroscientists constructed a large collection of images taken around Groningen, in the Netherlands (van Hateren & van der Schaaf 1990). They did some preliminary statistical analysis on these images. Later Lee et al. (2003) initiated a further study of this collection of images by constructing a dataset of 3×3 square patches of pixels. The idea was to study the statistics of very small patches, since such small patches are what would be expected to correspond to the single neuron

responses in V1. Starting with the 4167 natural images collected by van Hateren &
van der Schaaf (1990), they sampled 5000 (3×3)-pixel patches from each image, and
kept the top 20% with the highest contrast. The resulting set M of 4 167 000 patches
was subsequently normalized to force the contrast-measuring D-norm and the Euclidean
norm both to take the value 1 on each patch. These normalizations project the data
set from the nine-dimensional space in which it starts out onto the surface of a seven-
dimensional sphere S^7.

While the resulting point cloud is dense on S^7 in the sense that all points on the
sphere are close to a point in the data set, the analysis by Lee et al. (2003) found that the
points concentrate in a high-density annulus, since then dubbed the *primary circle*, as
seen in Figure 6.1. In the paper of Lee et al., a coordinate change for $\mathbb{R}^8$ was given that
concentrates this primary circle into the first two coordinates. The shaded squares in the
figure are continuous approximations to the 3×3 patches, using gray-scale values. Using
the data set constructed by Lee et al. Carlsson et al. (2008) analyzed the more subtle
aspects of the geometry and topology of the data set. In this chapter, we will walk through
the constructions and arguments, and illustrate their observations on the data set.

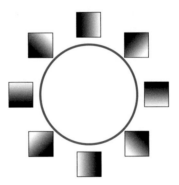

Figure 6.1 The primary circle

In order to fully understand this data set, we need first to understand what is meant
by *high-density subsets*. There are two components to this: on the one hand, what do
we mean by *density*, and on the other hand, what do we mean by *high*. For our present
considerations, we will use the density proxy which we called the codensity, defined
in Section 2.8, and which we denote by δ_k. The parameter k in the definition is a
threshold, where we define $\delta_k(x)$ as the kth nearest neighbor of a point. For any k,
we can evaluate δ_k on points of the Mumford–Lee–Pedersen data set M. Once that is
computed, we can introduce a second parameter τ, which is a percentage. By a set of the
type $M_0(k, \tau)$, we mean a sample of 50 000 points sampled from the points of M that
are among the τ percent highest points in δ_k-value. We then construct a witness complex
$W_{50}M_0(300, 30)$ with 50 landmark points on $M_0(300, 30)$. Computing its persistent
homology, we find a degree-1 barcode as in Figure 6.2.

We can vary the sample and the choice of 50 landmark points, and we always obtain the
same behavior as seen in Figure 6.2: many small and short intervals and one large, long,

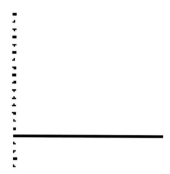

Figure 6.2 The β_1-barcode for $\mathcal{W}_{50}\mathcal{M}_0(300, 30)$. Here β_1 is the Betti-1 number, i.e. the dimension of the first homology, H_1

interval corresponding to the primary circle. That the long interval really corresponds to the primary circle may be established by computing a representative cycle for the homology class and sampling points along the cycle. Performing this sampling gives patches that exhibit the linear gradient behavior that we expect.

As we saw in Figure 2.3, decreasing the parameter k corresponds to capturing finer features of the data and smoothing it less. We notice on the left-hand side of Figure 2.3 that the points picked out show up wherever a small dense cluster is formed, while the points on the right-hand side pick out the central cluster of the entire data set. Thus, we may well expect that witness complexes on the Mumford–Lee–Pedersen data set with a lower k-value would pick out more subtle features. Indeed, the β_1-barcode for $\mathcal{W}_{50}\mathcal{M}_0(15, 30)$, as can be seen in Figure 6.3, picks out five circles in the data set. Again, the resulting barcode is stable, in that it reappears for many different randomly selected witness point families.

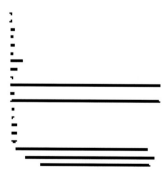

Figure 6.3 The β_1-barcode for $\mathcal{W}_{50}\mathcal{M}_0(15, 30)$

This barcode can certainly be realized by a bouquet of five circles, as on the left in Figure 6.4. However, there are other shapes that realize the five loops. By a combination of examination of the data and mathematical experimentation with different models, one can arrive at the model on the right in Figure 6.4 as another possibility.

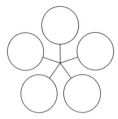 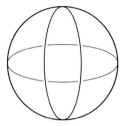

Figure 6.4 Spaces with $\beta_0 = 1$ and $\beta_1 = 5$. Both spaces are homotopy equivalent to a bouquet of five circles; however the arrangement on the right has an interpretation in the data.

We can find interpretations for the different components of the three-circle model on the right in Figure 6.4. Consider the layout in Figure 6.5. The red circle is the primary circle that we have already seen in Figure 6.1. The green and yellow secondary circles, however, may be interpreted as corresponding to horizontal and vertical non-linear gradient families, respectively, giving, in the end, a situation as in Figure 6.6.

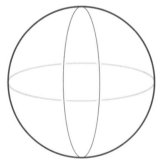

Figure 6.5 Three-circle model

In this interpretation, the primary circle captures linear gradients that change directions along the circle, while the two secondary circles capture vertically and horizontally aligned gradients that may be quadratic, in the sense that they may be ridges or valleys in a given orientation. Again, by sampling patches from the points involved in representatives for the corresponding homology classes, we may observe this behavior in the data and thus verify our interpretation.

It is interesting that the two first structures to appear after the primary circle consist of horizontal and vertical features. One can easily identify two possible explanations for this phenomenon.

1. Vertical and horizontal structures to occur more frequently in nature for stability reasons. Skew structures lack support and thus will fall over unless counterbalanced by something, whereas straight structures will remain standing without support.
2. The rectangular sensor grid of the camera biases the pictures captured so that they tend to align with the sensor grid that detects the pictures.

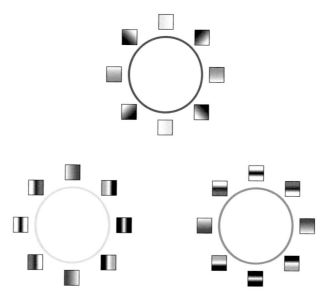

Figure 6.6 The three-circle model in the Mumford data

A similar analysis on the space of (5×5)-pixel patches from natural images reveals the same three-circle structure, however; this would be an argument in favour of explanation 1. Experiments suggest that both explanations contribute.

We recall from Example 3.89 that there is a space called the *Klein bottle*, which is defined using a simple equivalence relation among the points of the boundary of a rectangle. There is a natural way to embed the three-circle model in a Klein bottle. Consider Figure 6.7: the primary circle runs along the Klein bottle, circling it twice, while the secondary circles run transversely, intersecting the primary circle each time it comes past them.

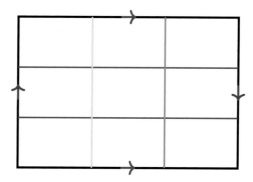

Figure 6.7 Three circles in a Klein bottle

However, the mere existence of an embedding of the observed model in a particular surface does not prove anything about the data. The next step of the argument is to demonstrate how the Klein bottle works as a non-linear parametrization of high-density high-variance patches. To do this, we first need a convenient model for the Klein bottle that is consistent with the pixel patches we are studying.

The Klein bottle occurs in many different mathematical contexts. A useful one for image applications is as a space of quadratic polynomials in two variables. The patches we have been looking at can be viewed as a discretization of such polynomials by restricting them to a 3×3 grid in the plane, and we obtain a subset of these polynomials which is homeomorphic to a Klein bottle and which will be consistent with our findings on image patches. Define Q to be the space of quadratic polynomials in two variables x, y. Since any such polynomial can be written as

$$f(x, y) = A + Bx + Cy + Dx^2 + Exy + Fy^2,$$

the resulting space Q is a six-dimensional real vector space. We note that any such polynomial can be seen to correspond to a picture underlying a pixel patch by viewing the polynomial as a function and returning intensities as defined on the unit square I^2, where I denotes the interval $[-1, 1]$. The corresponding image patch is obtained by restricting the polynomials to the set $\{-1, 0, 1\} \times \{-1, 0, 1\} \subseteq I^2$. The normalization steps performed by Lee et al. (2003) translate as follows. The first step was mean-centering, which carries a patch to a corresponding patch whose mean value is equal to zero. The corresponding construction for functions carries any quadratic polynomial function f to the function

$$\hat{f} = f - \frac{1}{4} \iint_{I^2} f \, dx \, dy.$$

We will denote the set of all $f \in Q$ for which $\iint_{I^2} f \, dx \, dy = 0$ by $\mathcal{P}$. The set $\mathcal{P}$ is a five-dimensional vector subspace of the six-dimensional vector space of all quadratic polynomials in two variables. We will also denote the set of all $f \in \mathcal{P}$ which are not identically zero by $\mathcal{P}^*$. This corresponds to the filtering step, which removes all patches which are too close to being the zero patch. Next, for any $f \in \mathcal{P}^*$, we can form the function $\overline{f} = \frac{f}{\|f\|_2}$ where

$$\|f\|_2 = \left(\iint_{I^2} f(x, y)^2 \, dx \, dy \right)^{1/2}.$$

Of course the function $\overline{f}$ satisfies the condition $\|\overline{f}\|_2 = 1$. We denote the subset of all $f \in \mathcal{P}^*$ satisfying $\|f\|_2 = 1$ by $\mathcal{P}_0^*$. As a space $\mathcal{P}_0^*$ is a four-dimensional ellipsoid within the original six-dimensional space of polynomials.

Within $\mathcal{P}$ there is a subspace $\mathcal{P}_1$ consisting of polynomials $f \in \mathcal{P}$ such that

$$f(x, y) = q(\lambda x + \mu y),$$

where q is a quadratic polynomial in a single variable and $\lambda^2 + \mu^2 = 1$. The space $\mathcal{P}_2 = \mathcal{P}_0^* \cap \mathcal{P}_1$ is a two-dimensional manifold contained in Q.

Claim 6.1 *The space $\mathcal{P}_2$ is homeomorphic to the Klein bottle.*

Proof We write $\mathcal{A}$ for the space of polynomials $q(t) = c_0 + c_1 t + c_2 t^2$ such that

$$\int_{-1}^{1} q(t)\, dt = 0 \quad \text{and} \quad \int_{-1}^{1} q(t)^2\, dt = 1.$$

Regarded as a subspace of $\mathbb{R}^3$, this space is an ellipse, and thus homeomorphic to a circle. For any unit vector $\vec{v} = (\lambda, \mu) \in \mathbb{R}^2$ and any $q \in \mathcal{A}$, we write $q_v : \mathbb{R}^2 \to \mathbb{R}$ for $q_v(\vec{w}) = q(\vec{v} \cdot \vec{w}) = q(\lambda x + \mu y)$, where $\vec{w} = (x, y)$. The function q_v clearly belongs to $\mathcal{P}_1$.
It is easy to check that

$$\int_{I^2} q_v = 0 \quad \text{and} \quad \int_{I^2} q_v^2 \neq 0$$

and thus

$$(q, \vec{v}) \mapsto \frac{q_v}{\|q_v\|_2}$$

defines a continuous map $\theta : \mathcal{A} \times S^1 \to \mathcal{P}_0$, where

$$\|q_v\|_2 = \int_{I^2} q_v^2.$$

We now observe that the map τ on one-variable polynomials given by

$$\tau(c_0 + c_1 t + c_2 t^2) = c_0 - c_1 t + c_2 t^2$$

restricts to give a map $\tau : \mathcal{A} \to \mathcal{A}$, and we note that $\tau \circ \tau = id_{\mathcal{A}}$. It is now evident that θ is not a homeomorphism, since $\theta(q, \vec{v}) = \theta(\tau(q), -\vec{v})$. Therefore, the map θ factors through a map $\overline{\theta} : \mathcal{A} \times S^1/\sim \to \mathcal{P}_2$, where $\sim$ is the equivalence relation given by $(q, \vec{v}) \simeq (\tau(q), -\vec{v})$.
We now show that the factorization $\overline{\theta}$ is a homeomorphism. We will prove this in three steps. The first is to observe that if $\theta(q, \vec{v}) = \theta(q', \vec{v})$, then

$$\frac{1}{\|q_v\|_2} q = \frac{1}{\|q'_v\|_2} q'.$$

For, $\theta(q, \vec{v}) = \theta(q', \vec{v})$ means that $Q \circ \sigma_v \equiv Q' \circ \sigma_v$, where

$$Q(t) = \frac{1}{\|q_v\|_2} q_v, \quad Q' = \frac{1}{\|q'_v\|_2} q'_v,$$

and $\sigma_v(x, y) = \lambda x + \mu y$. The map $\sigma_v : \mathbb{R}^2 \to \mathbb{R}$ is surjective, and it follows that $Q(t) = Q'(t)$ for all t, which means that Q and Q' are equal as functions. In particular, q and q' differ by a positive constant multiple. Since we are given that

$$\int_{-1}^{1} q(t)^2 dt = \int_{-1}^{1} (q'(t))^2 dt = 1,$$

it follows that $q = q'$.
The second step is to observe that if $\theta(q, \vec{v}) = \theta(q', \vec{v}')$, then $\vec{v} = \pm \vec{v}'$. To see this, we evaluate the Jacobian matrices $J_{(x,y)} \theta(q, \vec{v})$ and $J_{(x,y)} \theta(q', \vec{v}')$. This calculation is performed using the chain rule. We find that

$$J_{(x,y)} \theta(q, \vec{v}) = \frac{1}{\|q_v\|_2} q(\sigma(x, y)) \cdot \vec{v} \quad \text{and} \quad J_{(x,y)} \theta(q', \vec{v}') = \frac{1}{\|q'_{v'}\|_2} q'(\sigma(x, y)) \cdot \vec{v}'.$$

This tells us that there is a number r such that $r\vec{v} = r\vec{v}'$. But since $\vec{v}$ and $\vec{v}'$ are both unit vectors, this tells us that $\vec{v} = \pm\vec{v}'$. Suppose now that $\theta(q, \vec{v}) = \theta(q', \vec{v}')$. It follows from the second step that $\vec{v} = \pm\vec{v}'$. If $\vec{v} = \vec{v}'$, then the first step tells us that $q = q'$. If, on the other hand, $\vec{v}' = -\vec{v}$, we observe that

$$\theta(q, \vec{v}) = \theta(q', -\vec{v}) = \theta(\tau(q'), \vec{v})$$

and that it therefore follows from the first step that $\tau(q') = q$, and consequently $q' = \tau(q)$. However, the pairs $(q, \vec{v})$ and $(\tau(q), -\vec{v})$ are identified in the quotient space, so we find that the map θ is a continuous bijection on points. That θ is a homeomorphism follows from Proposition 3.86.   □

Merely experimenting with the density proxy δ_k, sample sizes, and resamplings is not quite enough to actually realize this model of the Klein bottle: the three circles are easy enough to exhibit, but a naïve approach fails to demonstrate a non-trivial β_2, as would have been expected for the homology of the Klein bottle with coefficients in $\mathbb{F}_2$.

In order to obtain an actual Klein bottle, we use the intuition that we have already obtained that there is a preference for vertical and horizontal patches. For that reason, we expect that patches that are instead in a direction which is 45° from the vertical or horizontal would occur much more rarely. Among the 45° patches, we would further expect that quadratic patches would be of much lower density. Accordingly, we compensate by including a small number of patches in the data set which are close to points that are not aligned in the horizontal or vertical directions, and moreover exhibit behavior close to pure quadratic.

These points form two pairs of one-dimensional arcs within the Klein bottle, as indicated in Figure 6.8.

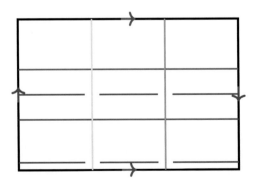

Figure 6.8 In blue, the four arcs needed for the Klein bottle. The image suggests that the endpoints of the lower of the blue arcs do not meet across the identification boundary, but we still intend that they should do so.

By adding points along the blue arcs to our sample from the data set $M(100, 10)$, we are able to find a witness complex with 50 landmark points on this enlarged sample set that display a non-vanishing β_2 as well as the expected β_1 and β_0 values, as seen in Figure 6.9.

The analysis suggests that the Klein bottle in the Mumford data set may be understood as in Figure 6.10. One of the major axes on the bottle is the orientation of the patch

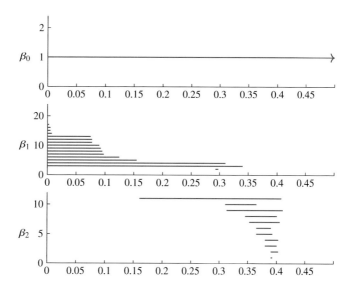

Figure 6.9 The barcodes for the amended point set exhibiting a Klein bottle structure in the Mumford–Lee–Pedersen data set. The vertical axis enumerates the bars and the horizontal axis corresponds to the filtration parameter.

gradient, while the other axis indicates which kind of gradient is drawn. This gives us the right kind of identification along the edges to motivate the Klein bottle we demonstrated in the data.

Remark 6.2   The procedure for adding additional points may be regarded as somewhat ad hoc, even though it is motivated by our understanding of the data obtained from the three-circle analysis. A methodology for making it more systematic is discussed in Brad Nelson's Stanford thesis (2020), under the heading of *parameterized topological data analysis*.

Remark 6.3   In the classification of surfaces in Hatcher (2002), it is shown that there are actually two two-dimensional surfaces that exhibit the same homology with coefficients in the field $\mathbb{F}_2$, one being the Klein bottle and the other being the torus. We have shown that what we are seeing is in fact consistent with a Klein bottle, using understanding gained from the data. We might want to be able to determine this without using the intuition obtained about the data, but in a more automatic way. It turns out that if one computes the homology with coefficients in $\mathbb{F}_3$ instead, that homology distinguishes between the torus and the Klein bottle. This computation was performed by Carlsson et al. (2008), and the result showed the Klein bottle.

Having this kind of description of the geometric structure of the data set can be useful in a number of ways.

1. **Compression schemes:** An understanding of what the frequently occurring data points are in a data set is often an important part of any compression scheme. For example, the *wedgelet* compression scheme (see Donoho 1999) uses the information

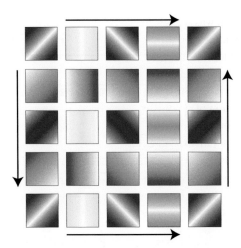

Figure 6.10 Representative pixel patches as distributed on the Klein bottle. The colors default to blue but are overridden to agree with the colors use in the three-circle model in Figure 6.6: patches that belong to the two "secondary" circles are colored yellow or green and patches that belong to the "primary" circle are colored red

that the primary circle in our description consists of frequently occurring patches in order to obtain an interesting compression scheme. In Maleki et al. (2008), a scheme based on the Klein bottle was constructed that outperformed wedgelets on some particular images. The rate distortion curve is given in Figure 6.11, where the Klein bottle compression scheme is the upper curve, and the lower curves are two different versions of the wedgelet scheme.

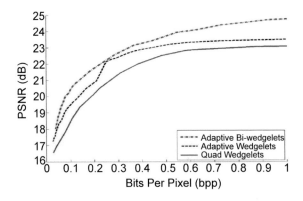

Figure 6.11 Rate distortion curve, describing the variation of the peak signal to noise ratio (PSNR) with bits per pixel, for Klein bottle-based compression. Reproduced from Maleki et al. (2008), with permission of IEEE, © 2008.

2. **Texture recognition:** In many image-processing situations, one is interested in recognizing not only large-scale features but also properties which are more related to the texture of regions of the image. For example, in studying textures, one expects to deal with the statistics of small features within the patch. One approach to this

problem was studied in Perea & Carlsson (2014). The idea was that one should should study all high-contrast patches occurring inside a (larger) texture patch statistically, with the hope that the statistics would tell the difference between textures. One way to understand the statistics is to locate, for each high-contrast patch, the closest point on the Klein bottle to that patch. If one carries out this process, one obtains a large collection of points on the Klein bottle. One can then generate a probability density function on the Klein bottle by smoothing. Since the Klein bottle has a very simple geometry (it has the torus as a twofold covering space), one can perform Fourier analysis on the probability density function to obtain coordinates which are the Fourier coefficients. It turns out that these coefficients are able to distinguish textures in various databases quite effectively, in fact close to state-of-the-art. The advantage of this method is that the effect of rotation on an image gives an easy transformation on the Klein bottle, namely translation in the horizontal direction in Figure 6.10. The standard approach to this problem is to identify a finite set of patches (the codebook), and to compute dot products of a given patch with each of the patches in the codebook, giving a distribution function on a finite set. The key observation here is that we can deal with infinite codebooks if the codebook is equipped with a useful geometry, as the Klein bottle is. Actual finiteness can be replaced by *finiteness of description*.

3. **Deep learning:** In Carlsson & Gabrielsson (2020) it was shown that the Klein bottle can be used as a tool for feature generation and for the design of topologically based deep learning architectures. Deep learning (see Goodfellow et al. 2016 and Aggarwal 2018) is a neural-network-based method for computations used in artificial intelligence. The foundation is a directed graph which specifies a computational pipeline, and it has been used very successfully in the analysis of data sets of images, as well as in many other domains. The discussion around Claim 6.1 above shows that points on the Klein bottle can be interpreted as functions on $[-1, 1] \times [-1, 1]$, and therefore by restriction to a 3×3 grid and, taking inner products, as a function or feature, on the set of patches. The idea of imposing a geometry on the features on data sets is the underlying philosophy of topological signal processing (Robinson 2014). It is useful in many ways, but one way in particular is that this kind of geometry suggests that graphs built from triangulations of the space in question (e.g. the Klein bottle) could lead to associated computation graphs which provide better performance, on various axes of comparison, for the study of various image data sets. This was demonstrated in Carlsson & Gabrielsson (2020) and Love et al. (2020).

6.2   Databases of Compounds

In the previous section, we saw that persistent homology can be a useful tool for understanding the overall structure of a complex data set. As we have pointed out, it is also possible to use persistent homology as a method of feature generation for complex unstructured data sets, where the individual data points carry a geometric structure. In this section, we will discuss a particularly useful example of this idea which is used for the study of organic compounds.

Databases of organic compounds are of fundamental importance in many biomedical applications. They are, for example, the main object of study in the area of drug discovery. Methods for interrogating them, and for determining rapidly and simply which compounds are functionally similar to which other compounds, are of critical importance in the development of useful drugs. In order to do this, one must assign notions of distance to pairs of compounds, where the distance function captures some notion of similarity. One difficulty is that sets of compounds do not fit neatly into standard formats for data, such as spreadsheets. The data which specifies a particular compound consists of position coordinates for the atoms which it comprises, as well as information about bonds between the atoms. Since different compounds consist of different numbers of atoms as well as different numbers of bonds, there is no simple vector description with a fixed number of coordinates which can describe a set of compounds. Even if there were the same number of atoms in all the molecules, the description would not be unique since a molecule could be rotated to produce different coordinates for the atoms. There is a standard method for encoding the structures of compounds, referred to as the *simplified molecular-input line-entry system*, or *SMILES* (see Weininger 1988). In this system, the compounds are represented as a list of symbols encoding the atoms, as well as other annotation symbols which permit the reconstruction of the compound. Three problems with this representation are that: (a) there is not always a unique SMILES representation of a given molecule; (b) the conversion from SMILES to a three-dimensional representation can be ambiguous; and (c) the structure is such that there is no obvious way of assigning a distance function or similarity score to a pair of SMILES structures. Another approach is to recognize that the molecules themselves can be thought of as finite metric spaces, with the atoms as points, and apply various forms of persistent homology to them to obtain persistence diagrams or barcodes. To complete the feature generation process, we then use the coordinatizations given in Section 5.2 to obtain numerical features characteristic of the molecule.

There are at least three ways to view a molecule as a metric space.

1. When the molecule is given as a list of atoms together with positions in Euclidean space, one can define a distance on the set of atoms as simply the Euclidean distance between the centers of the atoms.

2. When the molecule is given as a list of atoms and a list of bonds between atoms, we observe that the set of atoms now form the vertices of a graph, where any two atoms are connected by an edge if and only if there is a bond between them. In this case, one can define the graph distance between a pair of atoms as the minimal number of edges occurring in an edge path connecting the two atoms.

3. When the molecule is given as a collection of atoms and a collection of bonds together with bond lengths, one can consider the set of atoms as the set of vertices in a *weighted* graph. There is now an edge path distance defined on a pair $\{x, y\}$ of atoms as the minimal length of an edge path from x to y, where the length of an edge path is the sum of the weights of the edges occurring in the edge path.

Although there are many situations where all of these metrics can be used, we note that metric 1 is most appropriate in situations where where one is studying the molecule physically, so that the Euclidean distances are actually critical. On the other hand, metrics 2 and 3 are more appropriate where one does not have access to physical distances, or where one is considering a molecule that has many different conformations (i.e. embeddings in Euclidean space) and one wishes to study such molecules in a way which is independent of their conformation.

For a given metric, it is also very useful to construct modified filtered complexes which can emphasize important properties which one wants to study. Here are some examples of this kind of approach.

1. If one is given a real-valued function on the atoms (i.e. vertices), one can fix the scale parameter and create functional persistence complexes as in Section 4.7.3. There are numerous functions one could study in this way, or example, the atomic mass of the atoms, the *partial charge* of the atoms, or a centrality measure on the metric space defined by the molecule.
2. The atoms in a molecule are divided into groups depending on which element they represent or on the element type, such as halogens, metals, etc. Having constructed the Rips complex on the entire molecule, one can consider the subcomplex for which a simplex exists if and only if its vertices all belong to a fixed group of atoms.
3. One can also modify the filtrations directly using various strategies. For instance, one could modify the filtration using a function representing density. Rather than using the distance directly, one could filter by modifying a scale parameter on the basis of density information for pairs of atoms.

The efficacy of this kind of approach is well illustrated by the work of Guowei Wei and collaborators (Xia et al. 2015; Cang et al. 2018; Cang & Wei 2017, 2020; Nguyen et al. 2020a, 2020b). One of the important thrusts of this work is enabling *drug repositioning*, which refers to the use of a drug which is known to be effective for the treatment of one disorder to treat a different disorder. This is very important because, for example, one would already have an analysis of the possible side-effects of the drug, and it could therefore be put into use more rapidly. In Nguyen et al. (2020a), the authors used their methods for drug repositioning related to the Covid-19 virus, and they also developed general methods for drug discovery, which have been successful in competitive "challenges", in Nguyen et al. (2020b). The results go beyond data analysis and develop deep learning tools for the use of persistent homology information.

6.3 Viral Evolution

The phylogenetic tree has become the standard model to capture the evolution of species since Darwin. A single tree, *The Tree of Life*, contains all species, alive and extinct, in a single structure. The revolution in genomic technologies in the last 20 years has led to an explosion of data and methods developed to infer tree structure on sets of sequences (Felsenstein 2004; Drummond et al. 2002). This phylogenetic model strictly models *clonal* evolution, when genetic material is obtained from a single lineage of ancestors. In this case mutations can occur and then be transmitted via replication of the genomic material from a single parent to the offspring. It is known that there are numerous other mechanisms for the transfer of genetic material from one organism to another, where the organisms may even belong to different species. Recombination of genomic material is common in many species. Examples across species are pervasive in nature, however, as hybrids in plants, horizontal gene transfer in bacteria, and fusions of genomes as happened with mitochondria and chloroplasts. It was argued in Doolittle (1999) that trees do not adequately model the full range of possible mechanisms for the transfer of genetic material; thereby, conflicting results are produced in the reconstruction of the *Tree of Life*. It turns out that all trees have vanishing homology groups, and that therefore homological methods might be used to provide evidence for the presence of horizontal evolutionary events within a collection of sequences or in their genetic history. This idea was proposed in Chan et al. (2013), where a dictionary between algebraic topology and evolutionary concepts was outlined. The power of homology to study genomic data was shown with examples coming from viral evolution.

Such data sets consist of genetic sequences, which are sequences from a four-element alphabet given by $\{A, G, C, T\}$, corresponding to the nucleotides adenine, guanine, cytosine, and thymine. One natural metric on the set of such sequences is the *Hamming distance*, which assigns to a pair of sequences $\{x_i\}_i$ and $\{y_i\}_i$, with $x_i, y_i \in \{A, G, C, T\}$, the number of values of i for which $x_i \neq y_i$. Equivalently, one can describe the Hamming distance as the minimal number of substitutions which have to be made in one sequence to obtain the other. There are many variants on this distance which take into account the rates at which the various substitutions occur and assign different numbers to each possible substitution. Several variations of these distances were studied in Chan et al. (2013), all giving similar results. At that point we may apply persistent homology techniques to such finite metric spaces, and obtain persistence barcodes.

Since what we are trying to do is to distinguish some of these metric spaces from trees, it is important to understand the persistent homology of spaces which are in some way "tree-like". Let Γ denote a weighted graph, i.e. the triple $(V(\Gamma), E(\Gamma), f_\Gamma)$, where $V(\Gamma)$ is a finite set, $E(\Gamma)$ is a subset of the collection of two-element subsets of $V(\Gamma)$, and $f_\Gamma : E(\Gamma) \to (0, +\infty)$ is a weighting function on the edges. By an *edge path* in Γ, we mean a sequence $\{v_0, v_1, \ldots, v_n\}$ of elements $v_i \in V(\Gamma)$ such that for every $0 \leq i \leq n-1$ we have that $\{v_i, v_{i+1}\}$ is an element of $E(\Gamma)$. For any edge path $\mathfrak{e} = \{v_0, v_1, \ldots, v_n\}$, the *length* of $\mathfrak{e}$, denoted by $\lambda(\mathfrak{e})$, is given by

$$\lambda(\mathfrak{e}) = \sum_{i=0}^{n-1} f_\Gamma(\{v_i, v_{i+1}\}).$$

The distance, $d_\Gamma(v, v')$, between any two vertices v and v' of Γ is given by

$$d_\Gamma(v, v') = \min_{\mathfrak{e}} \lambda(\mathfrak{e}),$$

where the minimum is taken over all edge paths $\mathfrak{e} = \{v_0, \ldots, v_n\}$ with the property that $v_0 = v$ and $v_n = v'$. We will denote the metric space $(V(\Gamma), d_\Gamma)$ by $\mathfrak{M}(\Gamma)$. We now have the following result concerning the positive-dimensional persistent barcodes of a metric space $\mathfrak{M}(\Gamma)$.

Proposition 6.4 *Let Γ be any weighted graph, and assume that the underlying undirected graph is a tree, i.e. it admits no cycles. Then the Vietoris–Rips persistence barcodes $\beta_i(\mathfrak{M}(\Gamma))$ are all trivial, i.e. they form the empty multiset of bars, whenever $i > 0$.*

Remark 6.5 This result uses in an essential way the result proved in Buneman (1974) that a metric space is of the form $\mathfrak{M}(\Gamma)$ if and only if it satisfies the so-called four-point condition, which asserts that a finite metric space (X, d) is of the form $\mathfrak{M}(\Gamma)$ for some weighted tree if and only if the condition

$$d(x, y) + d(z, t) \leq \max(d(x, z) + d(y, t), d(x, t) + d(y, z))$$

holds for all quadruples of elements $x, y, z, t \in X$.

The effect of this result is that one can determine that a finite metric space is not tree-like, or that it is isometric to a finite subset of a space of the form $\mathfrak{M}(\Gamma)$. This statement alone is not useful in studying data, since there will be no "exactly tree-like" metric spaces occurring in the presence of noise. However, we have Theorem 5.2, which shows us that if a finite metric space is approximately tree-like, i.e. close to a tree-like metric space in the Gromov–Hausdorff metric, then its higher-dimensional barcodes will be close to the empty barcode in the bottleneck distance. This means that all the intervals contained in the barcode are of small length.

In Chan et al. (2013), a persistent homology analysis in dimensions 1 and 2 was applied to data sets arising in viral evolution, using the Vietoris–Rips complex from Section 4.3 of that publication and the witness complex defined in Section 4.3.4. While the details can be found in Chan et al. (2013), a summary of the results is as follows:

1. The type of exchange of genomic material can be catalogued by the topology of the genomic data. In the case of strict clonal evolution, all information is captured by the zero-dimensional homology. Segmented viruses, i.e. viruses whose genomic information is encoded in different "chromosomes" or segments, undergo reassortments: the generation of novel viruses with combinations of segments from the different parental strains. This phenomenon is the source of most of the reported influenza pandemics in humans (the H2N2 pandemic in 1957 and H3N2 in 1968), where a novel viral strain is generated by reassorting the reassortment of segments of different parental strains infecting different hosts. A data set of sequences arising in avian influenza was studied from this point of view. When each distinct segment was studied, there was no higher-dimensional behavior. However, the metric space produced by using the concatenated sequences did produce a significant one-dimensional homology, which would preclude phylogeny as a complete explanation. Other viruses, as HIV, the cause of AIDS, undergo recombination where the genome of the offspring is a mosaic of the genome of the parents. Circulating recombinant forms (CRFs) are examples of HIV viruses with complex recombinant patterns. Persistent homology shows higher-dimensional homology groups capturing some of the complex recombinant structures.
2. Estimation of the rate and scale of horizontal evolution: By performing persistent homological calculations on simulations, it was observed that a lower bound for the

recombination/reassortment rate was obtained by studying the counts of the number of bars in the corresponding barcode. By analyzing the numbers of these events it was possible to assess the evolutionary pressures that link different pieces of genomic material together. For instance, when analyzing the homology of avian influenza A, it was shown that several segments, containing the genes that encode the polymerases, are more likely to co-segregate together, possibly indicating that natural selection does not allow all combinations to be equally fit.

3. Higher-dimensional homology generators capture genomic exchange: In cases where higher-dimensional homology was observed, a representative cycle for a given feature (or bar) was obtained. Such a cycle is a linear combination of pairs of landmark points in the case of one-dimensional homology, of triples of landmark points in the case of two-dimensional homology, etc. The cycle then gives a list of data points that occur and this list can be investigated. In the cases that were studied, these occurrences were consistent with the known horizontal transfer mechanisms in these situations. It should be pointed out that the representative cycle was chosen directly from the algorithms used to compute the persistent homology, and no effort was made to find the minimal cycle, i.e. the cycle with the smallest number of summands in the formal sum. Choosing the minimal cycle would be expected to give even more focused outcomes. One might also attempt to reconcile the findings of homology with results obtained via the mapping methods introduced in Singh et al. (2007).

4. Explicit examples were studied: The triple reassortant avian virus that led to the outbreak in China in March 2013, the rate of reassortment of influenza A viruses infecting different hosts (humans, swine and birds), HIV recombination, and many other viruses including dengue, hepatitis C virus, West Nile virus and rabies.

In Lesnick et al. (2019), a systematic study of these ideas was carried out which pointed the way to a good mathematical theory.

6.4   Time Series

Time series form a very interesting class of data, where for a discrete variable t we have a point x_t in a metric space (X, d). Usually X is a Euclidean space $\mathbb{R}^n$. Given such a time series and a positive integer l, we can construct a data set out of the time series $\{x_t\}_t$ as follows. Let $\mathfrak{T}_l(\{x_t\}_t)$ denote the set of all fragments $(x_{t_i}, x_{t_i+1}, \ldots, x_{t_i+l})$ for every possible initial time t_i. From the distance function d on X, we can define a a distance function d_T on $\mathfrak{T}_l(\{x_t\}_t)$ by the formula

$$d_T((x_{t_i}, x_{t_i+1}, \ldots, x_{t_i+l}), (x_{t_j}, x_{t_j+1}, \ldots, x_{t_j+l})) = \left(\sum_{s=0}^{s=l} d(x_{t_i+s}, x_{t_j+s})^2 \right)^{1/2}.$$

To give an idea of how this works, we consider a time series given by

$$x_t = \sin(t\epsilon),$$

where ϵ is a very small threshold. If $l = 0$, then the data set we obtain is just the range of the sine function restricted to the set of multiples of the threshold ϵ. If ϵ is sufficiently

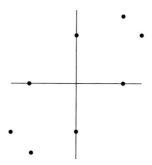

Figure 6.12 The time series when $l = 1$, $\epsilon = \pi/4$

small, it will roughly fill out the interval $[-1, 1]$, the range of the sine function. On the other hand, if $l = 1$ then the data points are pairs $(\sin(t\epsilon), \sin((t + 1)\epsilon))$. A diagram of this data set in the case where $\epsilon = \pi/4$ is given in Figure 6.12.

It is easy to check, using trigonometric identities, that this set is a discrete set of points lying on the ellipse

$$x^2 - \frac{\sqrt{2}}{2}xy + y^2 = \frac{1}{2}.$$

The fact that this data set contains a loop reflects the fact that the sine function is periodic. If we performed the same construction for a time series of the form $x_t = t\epsilon$, we would instead obtain a set of points sampled from the line with equation $y = x + \epsilon$. This set has no loops, which reflects the fact that the linear function $f(x) = x\epsilon$ is not periodic.

These observations suggest that the one-dimensional persistent homology or cohomology groups of the metric spaces $\mathfrak{I}_l(\{x_t\}_t)$ could suggest a strategy for detecting periodic behavior in a given function.

For low-dimensional signals, a delay embedding, as proposed by Takens (1981), mapping the time series a_t to a higher-dimensional $(d + 1)$-dimensional time series $(a_t, a_{t+\epsilon}, \ldots, a_{t+d\epsilon})$ produces a geometric shape. Takens proved that for appropriate choices of d and ϵ, this is an embedding that will recover the original dynamics from which the time series was measured.

This approach was explored by de Silva et al. (2012), Berwald et al. (2014), and Vejdemo-Johansson et al. (2015) using cohomology and circular coordinates as described in Section 4.5.6, and by Perea & Harer (2015) using persistent homology. We will describe both approaches here.

6.4.1   Intrinsic Phase Coordinates

For an approximately periodic function, the fact observed above that the point cloud will have a non-trivial degree-1 homology, and in particular a non-trivial degree-1 cohomology, means that we can extract coordinate functions generated by the geometry of the point cloud.

This idea was explored in de Silva et al. (2011), where persistent cohomology as described in Section 4.5.6 was defined. The connection between persistent cohomology and phase coordinates, i.e. the values that describe the progression through the period of a periodic function, comes from a result in classical algebraic topology which states that S^1 is the classifying space for the functor $H^1(-; \mathbb{Z})$, the integral version of cohomology defined in Remarks 3.125 and 3.126.

What this last sentence means is that homotopy classes of maps $X \rightarrow S^1$ correspond exactly to cohomology classes with integer coefficients from $H^1(X; \mathbb{Z})$. The correspondence is easy and constructive: we pick a simplicial complex X and a cohomology class $[z]$ with a representative cocycle $z \in C^1$. This means that $z: C_1 \rightarrow \mathbb{Z}$ is a map from the edges of X to integers and that $\delta z = 0$, that is to say $z(\partial y) = 0$ for all $y \in C_2$.

This last condition is a path-independence condition: the sum of the coefficients of the cocycle along paths should not depend on the path taken among homotopic paths. We may obtain a different result if we go around a hole, but as long as one path can be transformed into the other, the accumulated value from the cocycle should not change.

To build a map to S^1 from this, we map each vertex of X to 0, and use $z(e)$ for an edge e as a winding number, dictating how many times e should wind around S^1. The sign determines a direction for winding. Because of the path independence, paths that are homotopic will wind an equal number of times around the circle, so that the resulting function is continuous.

As it stands, this function is not very useful for real data, since application of it to a Vietoris–Rips complex maps each data point (i.e. each vertex in the Vietoris–Rips complex) to the origin in the circle. When it is described as a winding number, the function takes integral values. In order to obtain a useful construction, we will extend the coefficients to $\mathbb{R}$, so that we can vary the function continuously. This means that we will permit functions that send data points to circle values, and we will be studying $H^1(X, \mathbb{R})$. It is argued in de Silva et al. (2011) that the L_2-norm

$$\|z\|_2 = \sum_{e \text{ edge}} z(e)^2 \tag{6.1}$$

provides a good measure of how non-smooth z is. We can search for equivalent $\mathbb{R}$-valued cohomology classes that maximize smoothness by minimizing $\|z\|_2$ through only modifying it by coboundaries, so that we stay within the same cohomology class.

The winding method to build circle-valued coordinates remains applicable, with the change that an edge no longer needs to wind all the way around the circle but might traverse a portion of it. To produce a circle-valued function, we then would need to pick a starting data point, assign the value 0 to that, and then traverse the graph, assigning to each data point the accumulated values along a path from this starting point. By doing this the data points are pulled out evenly around the circle as well. The computation of this smoothest possible cocycle then corresponds to a least squares optimization problem

$$\min_{w \in C^0} \|z + \delta w\|_2. \tag{6.2}$$

The present authors discovered that, rather than explicitly performing the graph traversal step, it is enough to take a w that minimizes (6.2) and retain only the fractional part of each vertex value: then a function $f_z \colon X \to S^1$, with S^1 interpreted as the interval $[0, 1]$ with identified endpoints, is given by

$$f_z(x) = w(x) \pmod{1.0} \qquad \text{where } w = \arg \min_{w \in C^0} \|z + \delta w\|_2. \qquad (6.3)$$

6.4.2 Motion Capture and Indicator Functions

The techniques described in Section 6.4.1 were used by Vejdemo-Johansson et al. (2015) to study motion capture data. The data were originally generated by putting bright markers on humans who performed various motions – walking, skipping, climbing, swimming, etc – in a studio equipped with cameras recording their motions from several directions. By tracing the movement of the markers, and using an abstract model of a human skeleton (see Figure 6.13a) defined by a network of limbs with joints having prescribed ranges of motion expressed as angles, a time series of the motion can be constructed. Each observation in the time series is one possible pose, expressed by the three-dimensional position and orientation of a reference point – usually the hip – and activation angles for all joints in the skeleton model.

Each such time series traverses a curve in this *pose space*, which for the data used by Vejdemo-Johansson et al. (2015) was approximately 70-dimensional. Recurrent motions – motions that return to the same position again – trace out closed loops in pose space, and if the motion has several different types of motion, it will trace out several distinct curves. Each such curve generates a separate cohomology class, and each cohomology class generates a separate coordinate function describing how that particular recurrence traces out in the data. For motions, such as walking, that include a change in spatial position this curve will stretch out to a spiral, where the coordinates describing the hip position change even as the pose returns to a previously occurring configuration. For the application of motion recognition the data can be projected, removing the spatial position coordinates outright from the computation.

This separation into distinct coordinate functions becomes dramatically noticeable when one looks at the motion 143:23 from the CMU motion-capture data set (CMU Graphics Lab 2012), containing a boxing motion. The motion-capture recording shows punches and jabs with both right and left fists alternating. To illustrate what the motion looks like we can project the pose space onto two dimensions using a dimension-reduction technique, for instance principal component analysis (PCA). For the boxer, this produces the curve in Figure 6.13b.

6.4.3 Persistent Homology over Sliding Windows

Perea & Harer (2015) uses an extension of the Takens embedding definition: for any real-valued function f on the real line, define the *sliding window embedding* attached to f, with parameters $M \in \mathbb{N}$ and $\tau \in (0, +\infty)$, and denoted by $\mathrm{SWr}_{M,\tau} f(t)$, by

$$\mathrm{SWr}_{M,\tau} f(t) = (f(t), f(t + \tau), \ldots, f(t + M\tau)).$$

The quantity $\mathrm{SWr}_{M,\tau} f$ is an $\mathbb{R}^{M+1}$-valued function on $\mathbb{R}$. By choosing a finite set of points $T \in [0, L]$ we obtain a point cloud attached to the function f, which we denote by

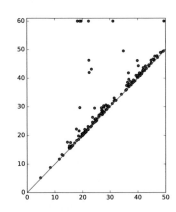

(a) A skeleton model of a human as used for motion capture.

(b) Projection of the curve described by the boxing moves in the motion recording 143:23.

(c) Persistence diagram for the boxing moves in the motion recording 143:23.

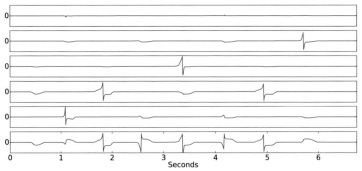

(d) The six circular coordinates remaining at parameter value 50 for the boxing moves in motion recording 143:23. Jointly they can be used to distinguish different specific motions from each other. ((a) reproduced from Vejdemo-Johansson et al. (2015), with permission of Springer–Nature, © 2015.)

Figure 6.13

$\mathfrak{C} = \mathfrak{C}(f, M, \tau, T)$. It is the persistent homology of $\mathfrak{C}$, using the Vietoris–Rips complex, that will reflect periodic behavior. To formulate the main result in Perea & Harer (2015), we make some definitions. For any L^2 function on the circle S^1, we let $S_n f$ denote its Nth Fourier truncation, i.e. the linear combinations of $\sin(nt)$ and $\cos(nt)$, for $n \leq N$, occurring in the Fourier decomposition for f. For any finite set $T \subseteq S^1$, we can then apply the sliding window embedding for $S_N f$, based on M and τ, and restrict it to T to obtain a point cloud in $(M + 1)$-dimensional Euclidean space $Y = Y(f, T, N, M, \tau)$. We then apply a process which mean-centers every data point in Y, and then normalizes it to lie on the unit sphere in $\mathbb{R}^{M+1}$. That is, for each data point $\underline{x} = (x_0, x_1, \ldots, x_M)$ we first transform it to a vector $\underline{\hat{x}} = (\hat{x}_0, \hat{x}_1, \ldots, \hat{x}_n)$, where

$$\hat{x}_i = x_i - \frac{1}{M+1}\left(\sum_i x_i\right).$$

Finally, we form $x^* = \underline{\hat{x}}/\|\underline{\hat{x}}\|$. We will denote a point cloud which has been transformed in this way by $\overline{Y}(f, T, N, M, \tau)$. We next assume that the function f is L-periodic on S^1, so that $f(x + 2\pi/L) = f(x)$. We then set

$$\tau_N = \frac{2\pi}{L(2N+1)}.$$

It was shown in Perea & Harer (2015) that the limit of the sequence $\mathrm{dgm}\,\overline{Y}(f, T, N, 2N, \tau_N)$ exists as a point $\mathrm{dgm}_\infty(f, T, w)$ in the completed barcode space $\hat{\mathfrak{B}}_\infty$ with respect to the bottleneck distance, defined in Section 5.3, where dgm denotes the operation which assigns to a point cloud its one-dimensional persistence diagram, or barcode, and where $w = \frac{2\pi}{L}$. For any barcode $\beta = \{(x_1, y_1), \ldots, (x_n, y_n)\}$ in $\mathfrak{B}_\infty$, we define its *maximal persistence*, denoted by $\mathrm{mp}(\beta) = \max_i(y_i - x_i)$. It is easy to check that mp extends to a function on $\hat{\mathfrak{B}}_\infty$. It is also proved in Perea & Harer (2015) that one may take a limit over families of subsets $T_\delta \subseteq S^1$, where T_δ is δ-dense in the sense that every point in S^1 is within a distance δ of some point in T_δ, and where $\delta \to 0$, to obtain the limit $\mathrm{dgm}_\infty(f, w)$. One of the principal results of Perea & Harer (2015) is the following.

Theorem 6.6 *Let f be an L-periodic continuous function on S^1, such that $\hat{f}(0) = 0$ and $\|f\|_2 = 1$, and suppose that the persistence diagram is computed for homology with coefficients in $\mathbb{Q}$. The we have that*

$$\mathrm{mp}(\mathrm{dgm}_\infty(f, w)) \geq 2\sqrt{2}\max_{n \in \mathbb{N}}|\hat{f}(n)|,$$

where $\hat{f}(n)$ is given by

$$\hat{f}(n) = \begin{cases} \frac{1}{2}a_n - \frac{i}{2}b_n & \text{if } n > 0, \\ \frac{1}{2}a_{-n} + \frac{i}{2}b_{-n} & \text{if } n < 0, \\ a_0 & \text{if } n = 0, \end{cases}$$

and where a_n and b_n denote the coefficients of $\cos(nt)$ and $\sin(nt)$, respectively, in the Fourier expansion of f.

This is a very interesting relationship, and these observations are used as a basis for a test for periodicity using persistence barcodes of the point clouds constructed from the function f. In Perea & Harer (2015) comparisons are made with more conventional methods for detecting periodicity, with favorable results.

6.5   Sensor Coverage and Evasion

6.5.1   The Coverage Problem

Questions of coverage arise in a variety of settings, such as for instance:

- Does a specific configuration of cell phone towers provide coverage for customers everywhere?
- How should navigational beacons be placed to best help robots navigate an area?
- For a more popular culture example, the plot of *Ocean's 8* (just like many other heist stories) relies on a failure of sensor coverage for the heist to succeed.

We can formalize the problem in the following way. Suppose that we have a collection of sensors contained in a domain $\mathcal{D}$, and that each sensor is able to sense the presence of an intruder or another sensor within a region containing it called its *footprint*. The question is whether these footprints cover $\mathcal{D}$. If that is the case, any intruder would be sensed by at least one sensor. If we had information about the locations of each sensor then it would be a relatively simple matter to determine whether the coverage condition holds. However, the assumption is that the sensors are very simple, and in particular do not have access to GPS or other methods for determining their positions. The only information they have available to them is the collection of sensors that are within their footprint, and the question is whether that is enough to determine whether the coverage condition holds. This problem was studied by de Silva & Ghrist (2006).

To understand their work, we first make an observation about a simple situation in singular homology.

Proposition 6.7   *Let D^2 denote the unit disc in the Euclidean plane, and let S^1 denote its boundary circle. Let $X \subseteq D^2$ denote any subspace containing S^1. We may consider the long exact sequence of the pair (X, S^1) from Proposition 3.143. The relevant fragment looks as follows:*

$$H_2(X) \longrightarrow H_2(X, S^1) \overset{\delta}{\longrightarrow} H_1(S^1).$$

Then $X = D^2$ if and only if the connecting homomorphism δ is non-zero.

Proof   We will use this result as an opportunity to illustrate the indirect methods of Section 3.3.10 as well as the method referred to as *commutative diagrams*. We will give the reasoning in full detail to illustrate all the points. In general there is a lot of shorthand which renders this proof as a two or three line argument.

One property of the long exact sequence of a pair that we did not mention in the statement of Proposition 3.143 is its naturality. That statement says that if we have two pairs of spaces (X_0, Y_0) and (X_1, Y_1) such that Y_i is a subspace of X_i for each i, and a continuous map $f : X_0 \to X_1$ such that $f(Y_0) \subseteq f(Y_1)$, then there is a commutative diagram

$$
\begin{array}{ccccc}
H_i(X_0) & \longrightarrow & H_i(X_0, Y_0) & \xrightarrow{\;\delta_0\;} & H_{i-1}(Y_0) \\
\downarrow & & \downarrow{\scriptstyle H_i(f)} & & \downarrow{\scriptstyle H_i(f|_{Y_0})} \\
H_i(X_1) & \longrightarrow & H_i(X_1, Y_1) & \xrightarrow{\;\delta_1\;} & H_{i-1}(Y_1).
\end{array}
\tag{6.4}
$$

The notion of a commutative diagram is ubiquitous in algebraic topology. It is vitally important in applications and computation within the subject. A diagram (of vector spaces) is a directed graph with each node corresponding to a vector space and each arrow corresponding to a linear transformation from the vector space corresponding to the tail node to the vector space corresponding to the head node. Any path in the graph from a node corresponding to a vector space V to a node corresponding to a vector space W gives rise to a linear transformation from V to W by composition of the linear transformations corresponding to the constituent edges of the path. The diagram is said to be *commutative* if every path in the diagram between two nodes gives rise to the same linear transformation. In particular, in diagram (6.4) above, we have that the linear transformation $H_{i-1}(f \mid_{Y_0}) \circ \delta_0$ is equal to the transformation $\delta_1 \circ H_i(F)$.

Suppose that X is not all of D^2, such that it is missing a point p in the interior of D^2. By a reparametrization of the disc, restricting to the identity on S^1, we may assume that p is the origin. The inclusion $S^1 \hookrightarrow D^2 - \{0\}$ is a homotopy equivalence, since $D^2 - \{0\}$ is homeomorphic to $S^1 \times (0, 1]$, using the polar coordinate representation. This means that the linear transformations $H_2(S^1) \to H_2(D^2 - \{0\})$ and $H_1(S^1) \to H_1(D^2 - \{0\})$ are both isomorphisms, and examination of the long exact sequence of the pair $(D^2 - \{0\}, S^1)$ gives

$$
H_2(S^1) \xrightarrow{\alpha} H_2(D^2 - \{0\}) \to H_2(D^2 - \{0\}, S^1) \to H_1(S^1) \xrightarrow{\beta} H_1(D^2 - \{0\}).
$$

The transformations α and β are isomorphisms as we observed above. It now follows from the numerical criterion in Proposition 3.139 that $H_2(D^2 - \{0\}, S^1)$ is the zero vector space.

We now have a further commutative diagram,

$$
\begin{array}{ccc}
H_2(X, S^1) & \xrightarrow{\;\delta_1\;} & H_1(S^1) \\
\downarrow{\scriptstyle \phi} & & \downarrow{\scriptstyle id} \\
H_2(D^2 - \{0\}, S^1) & \xrightarrow{\;\delta_2\;} & H_1(S^1) \\
\downarrow{\scriptstyle \psi} & & \downarrow{\scriptstyle id} \\
H_2(D^2, S^1) & \xrightarrow{\;\delta_3\;} & H_1(S^1).
\end{array}
$$

Since the diagram commutes, we have that the transformation $\delta_1 : H_2(X, S^1) \to H_1(S^1)$ is equal to the composite $\delta_3 \circ \psi \circ \phi$. But this composite factors through the vector space

$H_2(D^2 - \{0\})$, which is the zero space from our discussion above. It follows that δ_1 is identically equal to zero. $\qquad\qquad\qquad\qquad\qquad\qquad\qquad\qquad\qquad\qquad\qquad$ □

We will apply this argument in the situation where X is the "covered region", i.e. the union of the footprints of all the sensors. We also assume that we have a collection of sensors lying on the boundary, called the *fence sensors*, and it will be assumed that the footprints of the fence sensors cover the boundary of the region. We are not able to deal with X directly, since we do not even know the positions of the footprints. However, we can obtain information about its homology by using the Čech complex, and the Vietoris–Rips complex and the relationship between them. We will consider the situation where the footprints are discs of fixed diameter. Here are the pertinent facts.

1. The footprints and therefore all their intersections are convex sets, hence contractible.
2. It follows from fact 1 that the nerve lemma 4.6 applies to the covering of the region covered by the footprints. Since the Čech complex for a threshold R is the nerve of the covering of X by balls of radius R, it follows under certain conditions on D (such as, for example, convexity) that the simplicial homology Čech complex is isomorphic to the singular homology of the covered region X. One can prove similar results for $H_i(X, \partial X)$, using the covering by the footprints of the fence sensors, under suitable conditions on the boundary.
3. The fact that the sensors can detect exactly when another sensor is within a distance R means that we are able to compute the Vietoris–Rips complex at threshhold level R for the set of sensors.
4. In Section 4.3.2, it was proved that in general there are inclusions

$$C^{\text{Cech}}(X, R) \subseteq \text{VR}(X, 2R) \subseteq C^{\text{Cech}}(X, 2R).$$

In this specific situation, i.e. for points embedded in the plane, the bound $2R$ is much too generous, and one can do much better.

De Silva and Ghrist considered the coverage problem for a region D which is homeomorphic to the unit disc with boundary. Let ∂D be the boundary of D. Let S denote the set of all sensors in D, and let S^∂ denote the sensors on the boundary ∂D. In de Silva & Ghrist (2006), an analysis of the Vietoris–Rips complexes $\text{VR}(S, R)$ and $\text{VR}(S^\partial, R)$ was made. They observed that there is a quotient chain complex $C_*(\text{VR}(S, R))/C_*(\text{VR}(S^\partial, R))$, and a connecting homeomorphism

$$\delta^{\text{VR}} : H_2(C_*(\text{VR}(S), R)/C_*(\text{VR}(S^\partial, R)) \rightarrow H_1(\text{VR}(S^\partial, R)).$$

They then determine criteria concerning the region D and the parameter R, such that the non-vanishing of the linear transformation δ^{VR} implies that the sensors cover the region D. This is now computable. Note that the result is only sufficient; one does not expect to find both necessary and sufficient conditions.

6.5.2   The Evasion Problem

It is also interesting to consider the situation where the sensors move with time. It turns out that it is quite possible to have moving configurations in which the sensors

Figure 6.14 The construction of the prism complex. From de Silva & Ghrist (2006). © 2006 by SAGE Publications, reprinted by permission of SAGE Publications, Inc.

never cover the region but in which it is nevertheless impossible for an intruder to avoid detection. For this reason, it is interesting to study a time-varying version of the coverage problem, where the question is whether an intruder can evade the sensors at all times, and we will call this the evasion problem. We can define the evasion problem as follows.

Given a set of sensors S_i in a region $\mathcal{D}$ that move in time, we obtain from each sensor a curve in $\sigma_i \colon \mathbb{R} \to \mathcal{D}$ in $\mathcal{D}$. The question is whether there is a curve $E \colon \mathbb{R} \to \mathcal{D}$ such that E is never within the footprint of any S_i. In other words, using the simplifications that we used for coverage above, is there an evasion path E such that $\mathrm{dist}(E(t), \sigma_i(t)) > r$ for all t?

Again, to make the problem more tractable, de Silva and Ghrist made some assumptions. In addition to the assumptions for the coverage problem mentioned above, they assumed that, while sensor nodes may come or go, fence nodes never move and never drop offline.

The construction of de Silva and Ghrist is somewhat complicated technically, and we give only an informal description. Stacking the Vietoris–Rips complexes at each sampled time step, and gluing them together using prisms connecting all simplices that remain in the next complex, they could track the passage of time topologically. In other words, in this *prism complex*, if $[v_0, \ldots, v_d] \in \mathrm{VR}_r S_*(t)$ and also $[v_0, \ldots, v_d] \in \mathrm{VR}_r S_*(t + 1)$, then insert $\Delta^d \times [0, 1]$ and glue $\Delta^d \times \{0\}$ to the simplex at time t and glue $\Delta^d \times \{1\}$ to the simplex at time $t + 1$. An illustration of the construction can be seen in Figure 6.14.

If we were to span a sheet across the entire prism that would catch an evader – even if the sheet billowed back and forth in the "time" direction. This intuition was captured by de Silva and Ghrist in their Theorem 7:

Theorem 6.8 *Given the listed assumptions, a continuous curve $p \colon [0, 1] \to \mathcal{D}$ must at some point t be underneath the cover, if there is some relative homology-class $[\alpha] \in H_2(\mathcal{P}VR_r S_*, \mathcal{F} \times [0, 1])$ such that the image of $\partial \alpha$ under the map induced on the homology by the projection $\mathcal{F} \times [0, 1] \to \mathcal{F}$ is non-zero.*

Adams & Carlsson (2015) later constructed a test similar to Theorem 6.8 which can be calculated in a streaming algorithm and which gives the same responses but with far less computational effort for large problems.

Instead of gluing the entire stacked complex into a single long prism complex, as constructed by de Silva and Ghrist, the zig-zag construction discussed in Section 4.8 works by using inclusion maps from the individual complexes to the prisms connecting adjacent pairs. The construction is illustrated in Figure 6.15.

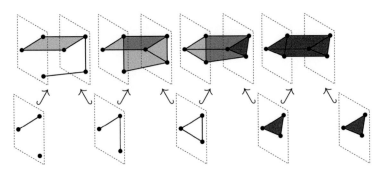

Figure 6.15 The zig-zag diagram for Adams and Carlsson's stacked complex. From Adams & Carlsson (2015). © 2015 by SAGE Publications, reprinted by permission of SAGE Publications, Inc.

As mentioned in Section 4.8, such zig-zag diagrams of vector spaces can be classified by barcodes, as can ordinary persistent homology. With this construction, one can show that if there is an evasion path in a domain then there is a full-length interval in the zig-zag barcode of homological degree 0. It turns out that the converse does not hold, since the zig-zag barcode can contain a full-length interval without corresponding to an actual evasion path. An example of this is given in Figure 6.16.

In other words, the Adams–Carlsson test, like the test provided by de Silva and Ghrist, is necessary but not sufficient.

The example in Figure 6.16 shows a sequence of sensor distributions with no α-witnessing coverage, but also no evasion path. Key to this example is that the evader cannot go backwards in time. If they could, an evasion path would exist in this example.

So far, we have been assuming minimal functionality from the sensors, i.e. communication and detection within known radii; however we have made no assumptions on detecting location, distance, or directionality of any kind.

There are indications that this amount of sensor capability is not sufficient to solve the evasion problem. Adams & Carlsson (2015) showed that for any domain in at least two dimensions, an example structurally similar to that shown in the lower part of Figure 6.16 can be found. Not only that, but the construction in the upper part of Figure 6.16 produces an example with identical Čech barcode but with an evasion path present. As long as the approach uses a complex approximating the Čech complex, these two constructions will appear identical. Adams and Carlsson concluded that, as a result, more sensor capabilities than just the homology data are needed to solve the evasion problem. More refined properties, such as the so-called *cup products* (see Hatcher 2002 and Carlsson & Filippenko 2020) can be used to obtain necessary and sufficient criteria for the existence of evasion paths, and there is even the potential for obtaining the structure of the space of evasion paths.

By increasing capabilities, Adams and Carlsson constructed a necessary and sufficient test for the existence of an evasion path. The extra capabilities they added are for the sensors to be able to detect cyclic ordering of neighbors; they are not necessarily able to quantify relative distances but they can keep track of the relative ordering of directions

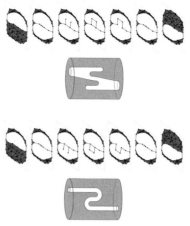

Figure 6.16 Upper: An example of where the de Silva and Ghrist criterion fails to recognize the lack of an evasion path. Lower: An example with an evasion path that for all currently available persistent homology constructions generates an identical barcode. The first and third rows in the figure represent covered regions at a collection of times, with time increasing as we move to the right. The grey objects in the second and fourth rows show the covered regions for all times, not just the seven time points displayed in rows one and three. The constructions in the odd rows will not be identical to those in the even rows, but are homotopy equivalent to them. From Adams & Carlsson (2015). © 2015 by SAGE Publications, reprinted by permission of SAGE Publications, Inc.

to other sensors. The capability increase provides the following theorem, which is Theorem 3 in Adams & Carlsson (2015).

Theorem 6.9 *Suppose we have a planar sensor network with covered region $X(t)$ connected at each time t. Then from the time-varying alpha complex and time-varying cyclic orderings of neighbors around each sensor we can determine whether an evasion path exists.*

The proof of Adams and Carlsson is highly constructive.

Proof The 1-skeleton of the alpha complex forms a planar graph. Doubling each edge into a pair of directed edges in opposite directions, these edges partition into directed cycles, each a boundary of a face in the planar graph. These cycles are locally determined by observing that each incoming edge is continued by its adjacent outgoing edge. An example is illustrated in Figure 6.17. As the alpha complex varies with time, regions may split, merge, and become small enough that they are covered by the sensor detection radii – all of which are moves that can be traced out using these edge partitions.

To track these changes, each boundary cycle is labeled with *true* if the corresponding face could potentially contain an intruder, and *false* if it cannot. At time $t = 0$, the outer boundary cycle as well as any length-3 cycles are labeled *false*, and all other cycles are labeled *true*. For each time step update, the following changes enumerate all possibilities.

1. If a single edge is added to the alpha complex, then a single boundary cycle splits into two. The two new boundary cycles inherit the evasion label from the boundary cycle that split.

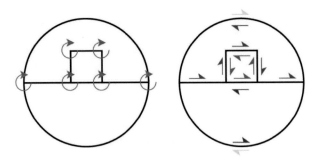

Figure 6.17 An example of the partition of the directed edges (right) into cycles bounding the faces of a planar graph (left). Each color corresponds to one cycle. From Adams & Carlsson (2015). © 2015 by SAGE Publications, reprinted by permission of SAGE Publications, Inc.

2. If a single edge is removed, the two boundary cycles of which its directed component edges are part are merged into a single cycle. This cycle is labeled *true* if either cycle was labeled *true*, and *false* only if both cycles were already labeled *false*.
3. If a single 2-simplex is added to the alpha complex, its boundary is set to *false*. When the 2-simplex is removed, the label remains.

An evasion path exists if and only if at the end of the time sequence there is a boundary cycle labeled *true*.   □

Khryashchev et al. (2020) observed two important features of Adams' and Carlsson's solution. On the one hand, the proof constitutes of an algorithm, which can be parallelized using the way in which Delaunay complexes depend on local information. On the other hand, a trace of a successful run through the algorithm can be used to construct an enumeration of possible (homotopy classes of) evasion paths. The construction goes as follows.

1. At the end of the trace (the completion of the algorithm), any region that is still *true* contains one possible homotopy class of evasion paths.
2. As we step through the algorithm trace log backwards, at any time step one of the following can happen to each *true* region:
   - It was unaffected by the change – the region has a unique predecessor in the earlier time step.
   - It was the result of a new α-edge splitting a region into two – the region has a unique predecessor in the earlier time step.
   - It was the result of a *true* region and a *false* region merging by the removal of an α-edge – the region has a unique *true* predecessor in the earlier time step.
   - It was the result of two *true* regions merging by the removal of an α-edge – the region has *two* different predecessors in the earlier time step, and any evasion path that has been traced to this stage gets doubled.
3. Passing through the entire log, all possible evasion paths are produced by tracing these changes.

--------------------- For the advanced reader ---------------------

The result from Adams and Carlsson using the existence of a full-length barcode was replicated using cellular sheaves by Curry (2013). This sheaf-theoretic reinterpretation sets the stage for the next results that we will address. Ghrist & Krishnan (2017) proved a necessary *and* sufficient criterion for the existence of an evasion path that works for arbitrary domain dimensions, as opposed to Adams' and Carlsson's criterion, which just works in the plane, since it relies on the structure of a planar graph.

To get to this point, Krishnan and Ghrist first developed a significant new theory for the (co)homology of parametrized spaces. Introducing a notion of *positive cohomology cones*, they proved an Alexander duality

$$^+H^{\dim E-q-1}C = {}^+H_q(E \setminus C).$$

Krishan and Ghrist used the superscript plus signs to indicate that this is a novel construction. From this duality it follows that evasion is possible if and only if $^+H^{\dim E-2}C \neq \emptyset$.

For details on this approach, we recommend Ghrist & Krishnan (2017).

6.6   Vectorization Methods and Machine Learning

As pointed out at the beginning of Section 4.5, a persistence diagram is a multiset. Therefore there is no predictable size to a persistence diagram. In fact, the number of bars may be as large as the number of simplices in the complex attached to the data set. This can make it difficult or awkward to use persistent homology in an analysis pipeline, since a vast majority of machine learning methods expect a fixed-size vector.[1]

A commonly seen approach is simply to take the N longest bars and use their lengths or their death–birth ratios (a log-scale type of bar length) to give a vector of length N describing the persistence diagram. At its simplest, $N = 1$, we get what could be thought of as a *persistence norm*, i.e. the length of the longest bar corresponds immediately to the bottleneck distance to the empty diagram. In this section, we will discuss a number of applications of vectorization methods.

6.6.1   Functional Summaries

In Section 5.2, we introduced several methods for vectorizing persistence bar codes, including symmetric polynomials, persistence landscapes, and persistence images. These form the first examples of a range of different functional summaries available as descriptors of persistent homology. We will now describe a few other approaches in active use in the literature. Starting with an application of persistence landscapes, we will continue with an application of Euler characteristic curves and finish up with a new construction from Chung & Lawson (2019) that generalizes a large variety of vectorizations into a common framework.

---

[1] An interesting direction of study could be how to feed a recurrent neural network with persistence diagram inputs.

Persistence Landscapes

In Kovacev-Nikolic et al. (2016), the authors studied the structural changes in the maltose-binding protein (MBP) as it performs biological functions. They worked with three-dimensional structures obtained from the Protein Data Bank (Bank 1971), and used the persistent landscapes that we described in Section 5.2.2 to automate the classification of protein configurations between open and closed – i.e. with or without connection to a ligand. Their work is illustrated in Figure 6.18.

The MBP is constructed from 370 amino acid residues, whose positions can be extracted from the protein data bank entries. Through dynamical distances (Kovacev-Nikolic et al. 2016), the authors generated a distance matrix for the amino acid residues. From the distance matrix they then generated a Vietoris–Rips complex and computed the persistence landscapes for the homologies H_0, H_1, and H_2.

The authors reported that training a support vector machine on the landscapes produced accurate and complete separation between open and closed configurations. They further reported using the total area supporting the landscape as their summary quantity X for statistical inference.

Calculating a two-sample t-statistic on the one hand for the mean support area of persistence landscapes in either category separately, they compared the result with the same calculation over permutations that break the separation into open and closed structures. Among all 1716 possible permutations *none* produced as extreme a t-value as the separation into open and closed proteins, which produced a p-value of 5.83×10^{-4} for the cases of zero- and one-dimensional homology. For H_2, 3.96% of the permutations produced at least as extreme t-values for the mean support area statistic.

Euler Characteristic Curves

In Richardson & Werman (2014), the authors introduced the Euler characteristic curve (ECC) for cubical complexes produced from bitmap images, and suggested highly efficient algorithms for its computation. Recall from Section 4.4.2 that the ECC is calculated from a parametrized topological space Y_x by defining as function values $ECC(x) = \chi(Y_x)$, the Euler characteristic of the corresponding topological space. For the bitmap image case, the ECC traces changes in the topology of sublevel sets of the image as a cutoff point sweeps through higher and higher values.

To apply the ECC to geometric recognition tasks, the authors suggested working with local curvatures of three-dimensional objects: points on the object are included in order of their local curvature, and the Euler characteristic is accumulated over this pointwise filtration of the objects. As a result, a curve is generated for each three-dimensional mesh; the curves have different geometric distributions of high- and low-curvature regions between different objects.

Two data sets are evaluated for three-dimensional recognition. In both cases, Richardson & Werman (2014) estimated the Hessian at each vertex. The eigenvalues $\kappa_1 > \kappa_2$ of the Hessian estimate the local curvature. The authors used these for two different filtration functions: the *maximum curvature*, $\max\{\kappa_1, \kappa_2\}$, and the *shape index*,

$$\frac{2}{\pi} \arctan \frac{\kappa_2 + \kappa_1}{\kappa_2 - \kappa_1}.$$

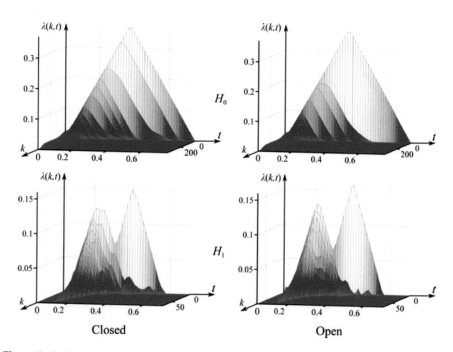

Closed   Open

Figure 6.18 Average persistence landscapes for MBP configurations. From Kovacev-Nikolic et al. (2016), republished with permission of Walter de Gruyter and Company, permission conveyed through Copyright Clearance Center, Inc.

Richardson and Werman used the ECC as a feature selection process for classification: the curves are used as input to a support vector machine classifier and evaluated with random four-fold and eight-fold cross validation; this was repeated 100 times.

First, they worked on the TOSCA (Bronstein et al. 2008) data set with high-resolution categorized triangulated meshes. An example can be seen in Figure 6.19. Using the shape index, a selection from the objects in the TOSCA data-set produces Euler characteristic curves as seen in Figure 6.21, where the separation of object groups can be clearly seen. The classification accuracy here – in recovering the categories from the data set – is high (as seen in Table 6.1): 98.6% using maximal curvature and 99.6% for the shape index.

Next Richardson and Werner examined the Lithic data set: a collection of scanned prehistoric stone tools, collected from two excavation sites, Qesem and Nahal Zihor (Grosman et al. 2008). An example is shown in Figure 6.20. As can be seen in Table 6.1, the two methods again achieve high precision: 93.7% for the maximum curvature method and 96.8% for the shape index.

Persistence Curves

A third approach is that of persistence curves, introduced by Chung & Lawson (2019). Persistence curves generalize both the rank invariant and the Euler characteristic curve into a joint framework. Explained with reference to the persistence diagram, a persistence curve is constructed by sliding a quarter-plane $Q_x = \{(x + \epsilon, x + \delta) \mid \epsilon, \delta > 0\}$ along

Figure 6.19 Sample cat from the TOSCA dataset. Left, the original cat. Middle, the maximum curvature function on the cat. Right, the regions of the mesh where the maximum curvature exceeds 0.05. From Richardson & Werman (2014), © 2014, with Permission from Elsevier.

Figure 6.20 A 3D-scanned stone tool from Nahal Zihor. Left, the original object. Middle, the maximum curvature function. Right, the regions where the maximum curvature exceeds 0.007. From Richardson & Werman (2014), © 2014, with Permission from Elsevier.

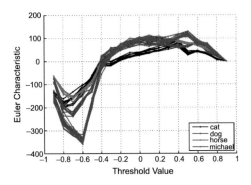

Figure 6.21 The ECC of the shape index filtration on TOSCA objects. From Richardson & Werman (2014), © 2014, with Permission from Elsevier.

Method	TOSCA (%)	Lithic (%)
ECC maximum curvature	98.6	93.7
ECC shape index	99.6	96.8

Table 6.1 Classification accuracies for Euler characteristic curves on the TOSCA and the Lithic data sets

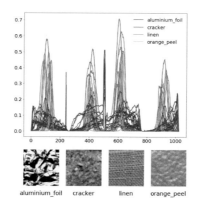

Figure 6.22 The upper part of the figure shows a collection of persistence curves, each from a separate texture sample, while the four lower images show the textures used. These persistence curves employ the sum of the lifetimes of the captured points as their summary functions. From Chung & Lawson (2019).

the diagonal (x, x) of the persistence diagram. For each sub-multiset $D \cap Q_x$ acquired through this process, the height of the curve at the corresponding coordinate x is given by applying some summary function f to the points contained in that quarter-plane. Some examples include:

- $f(D \cap Q_x) = \#\{D \cap Q_x\}$ – the rank function.
- $f(D \cap Q_x) = \sum(-1)^d \#\{D^{(d)} \cap Q_x\}$, where $D^{(d)}$ is the persistence diagram of the d-dimensional classes – the Euler characteristic curve.
- $f(D \cap Q_x) = \sum_{(b,d) \in D \cap Q_x} (d - b)$ – the curves depicted in Figure 6.22.

6.6.2 Vectorization in Action

In more recent years, the use of persistent homology for feature generation as input to other machine learning processes has taken root, with several designs that connect persistent homology with deep learning.

TopoResNet

In Chung et al. (2019), the authors studied a data set with photographs of skin lesions, by augmenting the deep learning architecture ResNet (He et al. 2016) with vectorized persistent homology features. As input to a deep learning network, they used two sets of persistent homology outputs on each of three color channels. Writing b for a birth and d for a death, forming a persistence interval (b, d), they used mid-life coordinates $M = (b + d)/2$ and lifespans $p = (d - b)/2$ to produce as statistical inputs, all taken separately in homological dimensions 0 and 1:

- the means of M and p;
- the standard deviations of M and p;
- the skewness of M and p;
- the kurtosis of M and p;
- the medians of M and p;

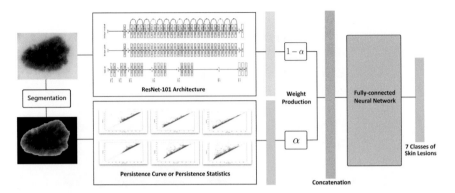

Figure 6.23 Architecture of the TopoResNet-101 architecture, where the skin lesion image goes through both the ResNet deep learning model as well as through cubical persistent homology to generate feature vectors that are weighted, concatenated, and classified using a final joint neural network layer. Reprinted, with permission, from Chung et al. (2019), © 2013, IEEE.

- the quartiles of M and p;
- the interquartile ranges of M and p.

These were combined with the persistence curves described in Section 6.6.1 above to form the input to an extension of the ResNet (He et al. 2016) architecture. The final architecture of the skin lesion classifier can be seen in Figure 6.23. The addition of topological features brings the classification accuracy up from 80.6% to 85.1% .

Protein Scoring

Cang & Wei (2018) studied protein structures using enriched versions of persistent homology as feature selection input to both k-nearest-neighbor methods and to a deep convolutional network setup. In their work, they used several different setups for both Vietoris–Rips and alpha complexes, as well as several different vectorization methods.

For vectorization, the authors used several approaches:

1. a computation of a family of barcode statistics, similarly to the way in which they are used in the previous example;
2. the statistics of binned persistence barcodes – dividing up the persistence intervals by birth, by death, or by lifetimes and calculating persistence statistics on each bin separately to generate longer feature vectors;
3. a two-dimensional representation that approximates representing multidimensional persistence as a discrete family of persistence diagrams, where all but one parameter is held fixed for each member of the family and persistence is calculated for the remaining parameter.

These vectorizations are used on persistence diagrams stemming from alpha complexes and from Vietoris–Rips complexes that are calculated in several ways:

Element specific PH Restricting to a subset of elements can focus the results on certain kinds of biological interactions.

Multi-level PH Editing the distance matrix to force disconnection between bonded atoms means that more subtle hydrogen bonds and van der Waals interactions can be emphasized.

Interactive PH Editing the distance matrix to force disconnection between all pairs of atoms except from between two chosen groups highlights interactions between these groups. This helps studying mutation sites or interactions between proteins and ligands, other proteins or nucleic acids.

Additional distances Correlation distance as well as flexibility and rigidity indices are used to construct distance matrices.

Electrostatic PH A distance derived from electrostatic interactions between charged particles is used to construct the filtration.

These choices of filtration functions and data subsets are combined into what Cang & Wei (2018) called *multi-component persistent homology*.

The results are fed either into k-nearest-neighbor regressors, gradient boosted tree regressors using bottleneck distances or Wasserstein metrics, or a convolutional neural network described in the paper. After testing combinations of these method choices on 4000 protein–ligand complexes and 128 374 compound-target pairs, they constructed an ensemble model *TopVS* using several of the combinations of the feature selections and complex constructions described here. Their ensemble model outperforms the state of the art when it comes to predicting binding affinity.

6.7   Caging Grasps

An important problems in robotics is how to instruct a robot to grasp an object. One approach is to instruct the robot to squeeze the object with enough force to compensate for the effect of gravity using friction to calibrate the required force. However, this requires quite a lot of knowledge and precision to find the right amount of force to use. If, for example, a robot is instructed to hold an apple in this way, it might easily do so with too much force, so that the apple would be damaged. *Caging grasps* form a different approach. A caging grasp does not work through balancing out friction and gravity with points of contact, but by creating a "cage", so that the grasped object can no longer move with complete freedom but is stopped by the robotic hand.

The difference can be seen with concrete, human relatable examples. When you hold up a playing card by pinching it between thumb and index finger, you are exerting enough force to overcome gravity and, by sensing the amount of pressure on your finger tips, you can keep the force at a comfortable but effective level. In contrast to this, when you lift a shopping bag, you will put fingers through the loops at the top and then move your entire hand up so that the bag hangs off your fingers. The bag is lifted because to avoid being lifted, it would have to pass through your fingers. By putting your thumb around the holders, you create a complete caging grasp: the grasp is not working through friction and pressure, but by restricting the range of possible motion for the object.

We can create a more abstract but mathematically clearer form of this example via the following formulation. Suppose that we have a loop $\mathcal{L}$ in ordinary three-dimensional space. We ask whether we can restrict the ways in which $\mathcal{L}$ can move by introducing an obstacle set O through which the loop is not permitted to move. If we have simple disc-shaped obstacles, it is easy to see that they do not materially obstruct the motion of $\mathcal{L}$. For example, $\mathcal{L}$ can move far away from any of the obstacles, so that all points of $\mathcal{L}$ will be far away from O. However, if O is itself a loop, and it links through $\mathcal{L}$, then $\mathcal{L}$ will not be able to move so as to be far away from O. If $\mathcal{L}$ moves so as not to cross O, there will always be a portion of $\mathcal{L}$ that remains close to O. What is interesting and relevant for this problem is the topology of the complement of O. One can see that the complement of a simple loop in $\mathbb{R}^3$ is homotopy equivalent to a circle S^1, and that its one-dimensional homology is therefore a single copy of $\Bbbk$. In order to create caging grasps for a more complicated object $\mathcal{L}$, it turns out to be useful to study $H_1(\mathcal{L})$. For example, consider a pretzel $\mathcal{P}$ in the shape of a figure eight. The complement of $\mathcal{P}$ turns out to have $\Bbbk \oplus \Bbbk$ as its one-dimensional homology. This result can be shown to be true using the *Alexander duality theorem* (see Hatcher 2002), which gives the structure of the homology of the complement of a subset $\mathcal{K} \subseteq \mathbb{R}^n$ in terms of the homology of $\mathcal{K}$ itself. The homology of the complement produces two ways of grasping the pretzel, namely by putting loops through one or the other of the two loops in it. The mathematical subject which studies the complements of various curves in $\mathbb{R}^3$ is referred to as *link theory*. A related question concerns the study of knots, which concern the distinct ways of embedding a loop in $\mathbb{R}^3$. A good starting reference for this subject is Rolfsen (1976).

One way that persistent homology enters the picture here is through viewing the sensor data of a robot as a point cloud X, and studying $H_1(X)$. The result of this computation lead to a collection of loops in X, and one can then define a caging grasp associated with each loop by constructing a loop in $\mathbb{R}^3$ which links with it. In general, homology classes in the complement of the point cloud in $\mathbb{R}^3$ will parametrize the different grasps.

Robots are nowadays often equipped with depth field cameras, such as can be found in a Kinect module for extending video games. When planning a grasp, we could assume that a robot has access to a point cloud sampled on the object it is trying to grasp, which is possibly incomplete and with noise.

In a sequence of three papers (Pokorny et al. 2013; Stork et al. 2013; Marzinotto et al. 2014), the robotics group at the Royal Institute of Technology in Stockholm developed the idea of using loops in these sensed point clouds to guide a caging grasp approach. To do this, we need *localized homology*, either using the methods described in Section 4.6, or algorithms from Erickson & Whittlesey (2005), Busaryev et al. (2010), and Dey et al. (2011) to calculate the persistent homology class having the shortest cycle representative. See Figure 6.24.

In the sensor data, we could treat any significantly long persistence H_1 class in the point cloud as representing a possible handle that can be grasped. By calculating a shortest-cycle representative, i.e. a cycle which has a minimal number of non-zero coefficients of simplices of the potential handle, we obtain a representation of the loop that is well localized, i.e. it contains as small a portion of the object as possible. The cycle traces out the handle while minimizing extraneous information. The details of the construction are complex, and are beyond the scope of this volume. However, we will give a quick description of the approach, with the expectation that the interested reader will consult the actual sources. To construct a control algorithm, one picks the following three constraints to cre-

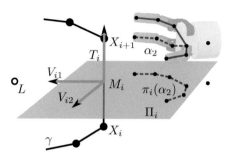

Figure 6.24 To evaluate a candidate edge T_i lying between X_i and X_{i+1} we construct a plane orthogonal to T_i, passing through its midpoint M_i. In this plane, the winding number of the projection $\pi_i(\alpha_2)$ of a path α_2 on the grasping hand around the origin at M_i is to be maximized. Reprinted, with permission, from Pokorny et al. (2013), © 2013, IEEE.

ate an optimization problem. Given a 1-class $[\gamma]$ with shortest-cycle representative γ and a particular edge $[\gamma_i, \gamma_{i+1}]$, the optimization is constrained as follows (see Figure 6.24).

1. Maximize the winding number of the projection of a model of the grasping hand to a plane orthogonal to the edge.
2. Minimize the distance between the grasping hand and the plane orthogonal to the edge that passes through the midpoint of that edge.
3. Avoid collision with the object.

The winding number can be calculated as

$$w(\gamma) = \frac{1}{2\pi} \sum \left[\tan^{-1} \left(\frac{\langle \gamma_{i+1}, \gamma_{i+1} - \gamma_i \rangle}{\gamma_{i,1}\gamma_{i+1,2} - \gamma_{i,2}\gamma_{i+1,1}} \right) + \tan^{-1} \left(\frac{\langle \gamma_i, \gamma_i - \gamma_{i+1} \rangle}{\gamma_{i,1}\gamma_{i+1,2} - \gamma_{i,2}\gamma_{i+1,1}} \right) \right].$$

Using data from a real sensor, the entire pipeline (the sequence of steps in the process) is illustrated in Figure 6.25.

The approach in Pokorny et al. (2013) was later refined in Stork et al. (2013). Again, take the example of picking up a bag by its handle. When aiming for the center of the minimal loop, the planning algorithm might end up trying to grasp around the entire bag rather than around the handle itself. As a solution to that, we could add a step that estimates the volume of the part of the loop that will be grasped, in order to distinguish a grasp around the bag from a grasp around the handle. The finished pipeline is depicted in Figure 6.26.

6.8 Structure of the Cosmic Web

A very interesting object of study in cosmology and astrophysics is the distribution of stars, galaxies, and other structures. Far from being uniform, it exhibits an intricate local structure, as indicated by the images in Figure 6.27.

Image (a) is from Huchra et al. (2005), image (b) is from NASA's Goddard Space Flight Center, and image (c) was provided by the European Southern Observatory and

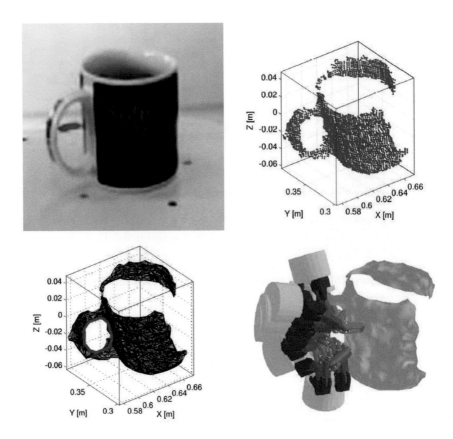

Figure 6.25 Sensors extract a point cloud (upper right) from a real-world object (upper left). With a filtered Delaunay mesh and a shortest-homology-loop computation, a topological representation (lower left) can be extracted and used to guide motion planning for a robotic grasping hand (lower right). The inferred surface is in mid-green and a collection of possible grasping poses are shown for a robotic hand. The arm is shown in pale green and the grasping fingers in red, pink, and blue – one color for each segment of each finger. Reprinted, with permission, from Pokorny et al. (2013), © 2013, IEEE.

Instituto de Astrofísica de Canarias. Galaxies and mass exist in a wispy web-like spatial arrangement consisting of dense compact clusters connected by elongated filaments and sheetlike walls, which surround large near-empty regions. See Bardeen et al. (1986) for a discussion of the relevant cosmology. To understand this kind of three-dimensional texture, one can attempt to study it as a topological problem. It is usually formulated via an assumption that the presence of matter is obtained by sampling from a density function ρ defined on three-dimensional space. The function ρ would itself be a complicated object, given the complexity of the resulting sampling. The first recognition that topological methods could be useful in this setting occurred in Gott et al. (1986) and Hamilton et al. (1986). The idea was that one should study the level sets of ρ from a topological point of view. These level sets are two-dimensional surfaces. For such surfaces there is a natural

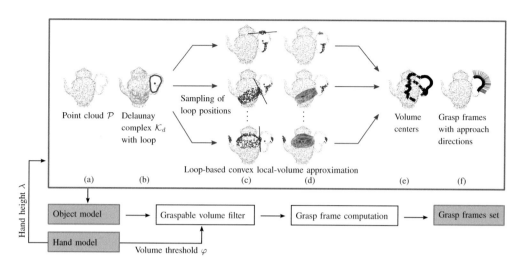

Figure 6.26 A full pipeline for planning grasps using minimal homology cycles in Delaunay complexes built from point data. The steps (c) and (d) help to remove grasp candidates that are not along a handle from consideration. Reprinted, with permission, from Stork et al. (2013), © 2013, IEEE.

(a) 2Mass redshift survey   (b) Goddard   (c) ESO and IAC

Figure 6.27 Distribution of structures in the universe: (a) from Huchra et al. (2005); (b) from NASA's Goddard Space Flight Center; (c) from the European Southern Observatory and Instituto de Astrofisica de Canarias

integer invariant called the *genus*, defined as one-half of its first Betti number. In Gott et al. (1986) and Hamilton et al. (1986), the genus of the level surfaces was considered as a useful numerical invariant of this situation. Note that the genus differs from the Euler characteristic, defined below, by a constant. Another family of spaces that one could study comprises the *excursion sets* at various levels r, defined as $\rho^{-1}([r, +\infty))$. It was recognized in Sousbie (2011) and Sousbie et al. (2011) that persistent homology gives a method which allows one to track the topological behavior of excursion sets as the thresholds for these excursion sets decrease.

In order to develop the theory, and to understand the appearance of the observed results, one can assume that ρ is itself a random function, or random field, obtained under some stochastic process. The aim is to understand the expected values for the Betti numbers of the excursion sets of random fields. Random fields were studied in great detail in Adler (1981) and Adler & Taylor (2007). A case of particular interest is that of a *Gaussian random field*, in which the assumption is made that the distribution assumed by the values of the random field at any particular point or any finite family of points in the domain is Gaussian. Gaussian fields in a cosmological context are of key importance because of the following facts.

1. The primordial universe was, to great precision, a spatial Gaussian random field; this has been observationally shown by numerous cosmic microwave background experiments such as COBE (Bennett et al. 2003), WMAP (Spergel et al. 2007), and more recently Planck (Abergel et al. 2011).
2. There is also a fundamental physical reason to expect that all primordial structure is nearly completely Gaussian. This concerns the generation of primordial density perturbations during the early inflationary phase of the universe (at $t \sim 10^{-36}$ seconds after the Big Bang), during which quantum fluctuations were blown up to macroscopic proportions.
3. A more technical mathematical reason why Gaussian fluctuations are expected is is the central limit theorem. Given the fact that the fluctuations at each scale are independently distributed, it would naturally give rise to a Gaussian field.

This assumption permits one to understand many properties of the random field, including some topological properties.

It turns out that for many Gaussian random fields (Adler & Taylor 2007) it is possible to determine the distribution of the Euler characteristic of the excursion sets, in terms of certain parameters defining the random field. These beautiful results are described in detail in Adler & Taylor (2007). An important case (when the Gaussian field is characterized by its *power spectrum*) was obtained in the cosmological literature (see Bardeen et al. 1986 and Hamilton et al. 1986). What this suggests is that one can begin to evaluate models for the stochastic processes from which the distribution in the cosmic web is generated, provided that one can develop sufficient theory to do so.

More recently, it was in Park et al. (2013) and van de Weygaert et al. (2011) that the Betti numbers of Gaussian random fields carry strictly more information concerning the random field than the Euler characteristic alone. Specifically, it was found that an invariant of the Gaussian field called the *slope of the power spectrum* affects the shape of the curves of the Betti numbers as the threshold r changes, while the shape of the curve of the Euler characteristics is independent of it.

Persistent homology has also been studied in this way. In Adler et al. (2010), the analogue of the Euler characteristic for persistent homology was defined and studied for Gaussian random fields. This analogue is defined as follows. For a fixed barcode $\beta = \{(x_1, y_1), \ldots, (x_n, y_n)\}$, we define the quantity

$$\tau(\beta) = \sum_i (y_i - x_i)$$

and then define the analogue of the Euler characteristic, χ^{pers}, by the formula

$$\chi^{\text{pers}}(X) = \sum_i (-1)^i \tau(\beta_i(X)),$$

where $\beta_i(X)$ denotes the i-dimensional persistence barcode for X. This quantity has the same properties as those which make the Euler characteristic highly computable, and in Adler et al. (2010) a result is proved which computes χ^{pers} for Gaussian random fields.

Finally, in van de Weygaert et al. (2011), a different topological approach was taken, to analyze the topology at various scales of a set of discrete points in two- or three-dimensional space. The points are to be thought of as individual stars, or perhaps galaxies, and the idea is to construct the α-complex (see Section 4.3.3) associated with them at various scales. This method was used in van de Weygaert et al. (2011) to study the Betti numbers produced by various stochastic-process models, including some which were explicitly proposed for cosmological problems. In addition, they pointed out that persistent homology is applicable to this family of complexes, and is likely to be applied in future investigations.

6.9   Politics

An explanation of the presence of a homology class is the *excluded middle*: data is present around a void but not in the middle of the void. One data set where we can see both this and the relative fragility of persistent homology to outliers is in the shape of voting patterns in the US House of Representatives.

We collected the voting results in all roll calls in the 2010 session of the US House of Representatives. For each vote cast, we encoded it as $+1$ if *yea*, -1 if *nay* and 0 for all other options. As a result we obtained a data matrix with 447 rows (one per representative in the data) and 664 columns (one per voting roll call). In Lum et al. (2013), this data was used to group congress representatives and find gradual and interconnected clusters (in a gradual cluster the degree of membership can vary, making the cluster look like a flare emerging from a core blob of data, as in the case of the diabetes data discussed earlier).

Another way in which the data matrix can be used, however, is by transposing it. In that way, each row represents a voting roll call, and the roll calls are represented by the sequence of votes that the representatives cast. Instead of investigating the shape of the collection of representatives in a space described by voting patterns on individual issues, this allows us to study the collection of issues in a space described by voting patterns of representatives.

Applying this idea to the voting dataset we obtained the following shape (after a PCA projection to two dimensions):

We see here a square shape, heavily populated in three of the four corners and somewhat more sparsely along the edges, and with a scattering of points in the middle. This turns out to be a consistent shape through the years. While the number of points in the middle and along the edges varies, the basic shape is remarkably stable.

The reason for this phenomenon emerges once we look at the properties of the data. Here, we have colored the points according to the sum of Democrat votes to the left, and the sum of Republican votes to the right; an explanation follows below.

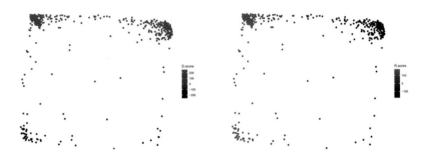

We used Ripser in Python (Tralie et al. 2018) to compute a persistence diagram (see below) on the 99% densest points in the data set.[2] In this diagram, showing us the persistence diagram for H_1, we can see a point lying noticeably off the diagonal. This point corresponds to the square-like shape visible in the PCA plots above and confirms to us that the structure in the PCA plots is present before the projection down to two dimensions.

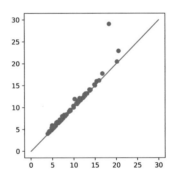

What are we seeing here?

Votes are concentrated in three of four corners of the square-like shape in the PCA plot. These corners correspond to full agreement from one or both parties, and the edges between them correspond to bills for which one party is unanimously agreed, but the other is split to a certain degree, with that degree depending on the position on the interval. The fourth corner would correspond to issues that are voted down strongly by both parties. Relatively few such issues are brought to the floor, but those that are appear to have somewhat greater density than those which have partial approval by both parties, which would correspond to the points in the center. It is interesting to inspect these bills. The top ten such, in density, include a group of bills that have no expectation of being passed but that are permitted to the floor. Examples would include resolutions to remove US troops from Afghanistan and Pakistan, as well as amendments to the Puerto Rico Democracy Act. Amendments to some other acts as well as motions to adjourn are also

---

[2] Using Gaussian kernel density estimation with bandwidth 1.0. We used Scikit-Learn 0.22.1 (Pedregosa et al. 2011) for the computation.

present. The reason for the general sparsity in the fourth corner is that without at least one party strongly backing a bill, it is very rare for that bill to come to a vote on the house floor. In the middle are issues that attract only parts of each party or that have a large number of votes recorded as 0 by the data collection script.

The homology structure that we can see here, which we would see even more clearly if we were to filter the data on density, is a result of the party whip structure. Most bills that reach the floor are championed by at least one party, so at least one party will vote nearly unanimously in favor. This places the bill on the boundary of the square spanned by two intervals: one for measuring the support from the Democrats, and the other for measuring support from the Republicans. The (relative) lack of points in the middle gives rise to a void that is detected by the persistent homology calculation. It indicates that it is more common for a bill to come to the floor with very little support from either party than it is for a bill to come to the floor with both parties having divided opinions on it.

As a comparison, we can perform exactly the same persistent homology computation but with all the data points – including the 1% that we had excluded in the computations above. When we do this, we get the following persistence diagram in dimension 1:

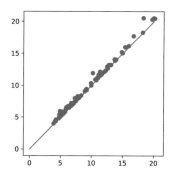

Where earlier we had a clear off-diagonal persistent homology class, here nothing distinguishes itself particularly from the noise on the diagonal. There is no significant structure to be seen. This is caused by the points in the middle of the square and in the lower right-hand corner. These outliers are removed when one is focusing on the 99% most dense data points.

For the diligent data analyst, the next question should be what these removed points have in common (if anything) with each other. Sitting in the middle of the square, these are bills that have partial support from each party. Digging in to the data, these were the roll calls 1, 54, 101, 424, 427 and 522 of the 2010 Congress session. Three of these are Quorum calls (Present/Not present) that got coded as 0 by the data collection script.

6.10 Amorphous Solids

Another interesting application of persistent homology occurs in materials science. There is a large body of theory around crystalline structures, which consist of very regular arrays of points in $\mathbb{R}^2$ or $\mathbb{R}^3$, in which the points represent atoms. They are regular in the sense that they are typically invariant under the action of translation groups in addition to certain other symmetries. The classification of such structures is highly developed. See Chatterjee (2008) or Wahab (2014) as references. We first show that Vietoris–Rips persistence barcodes applied to the collection of points viewed as a point cloud can in some cases distinguish between two different planar crystalline structures, one associated with a rectangular lattice in the plane and the other associated with a hexagonal lattice.

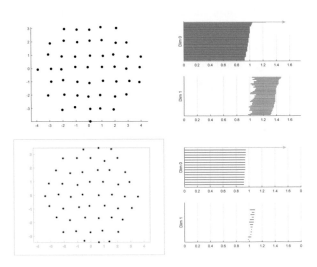

Figure 6.28 Slightly perturbed lattices and corresponding persistence barcodes. For each lattice the upper of the two figures on the right gives the zero-dimensional homology and the lower figure give the one-dimensional homology. The reason for the stark difference between the rectangular lattice homologies and the hexagonal lattice homologies is that we are using Vietoris–Rips complexes: the rectangular lattice produces rectangles and the hexagonal lattice produces triangles. The rectangles show up as homology classes while the triangles do not.

In Figure 6.28, the left-hand column shows segments of a rectangular lattice (upper) and a hexagonal lattice (lower), together with the corresponding persistence barcodes in the right-hand column. The lattices have been perturbed slightly to demonstrate robustness to noise. Note that the one-dimensional barcodes are quite different in the two cases, the bars in the rectangular case being much longer than those in the hexagonal case. It is interesting to ask if persistence barcodes can distinguish between all three-dimensional lattices. The authors guess that this is possible.

There are, however, many solids that do not exhibit such regular structure, and they are referred to as *amorphous*. They include glasses, rubber, gels, and many other solids. Although they do not exhibit a long-range regular structure, materials scientists detect certain kinds of less formal geometric patterns which distinguish between classes of such solids. These patterns are difficult to characterize and analyze, but material scientists have developed numerous ad hoc methods for studying them.

As a first attempt, one can study the short-range order, i.e. the statistics of pairs of nearest neighbors in the structure. This idea is important, but is insufficient in many situations, which motivates the study of longer-scale interactions, referred to as the *medium-range order*. A step in this direction is to study the distributions of bond angles and dihedral angles, which gives additional information but still only about the atomic configurations involving the second and third nearest neighbors. Another observed phenomenon that one wants to capture is the presence of hierarchical structures in the data. Such phenomena also cannot be studied with purely local statistics. In order to study such longer-scale interactions, one may consider the material as a network, with atoms as nodes and bonds as edges, and perform counts of rings (closed edge paths in the network) of various lengths and types. This approach gives yet more information, but it is only applicable to crystalline materials and other models where one can impose a network structure. Such a structure is not part of the data, which consists only of point positions, and therefore must be constructed by a method chosen by the investigator. Additionally, it does not take distances (bond lengths) into account and furthermore cannot account for hierarchical structures.

Persistent homology provides a systematic method which enables the study of geometric features such as rings without the imposition of a network structure and which can account for lengths and hierarchical structure. The material is studied by treating it as a point cloud in its own right, whose points are the atoms and where the distances are given by three-dimensional Euclidean distances. The persistence diagrams of such point clouds were used in Hiraoka et al. (2016) to quantify and make precise the geometric properties of several substances, namely silica glass, the Leonard–Jones system, and Cu–Zr metallic glass. We describe their results for silica glass.

Glasses are materials that are characterized by the fact that they undergo a "glass transition" when heated. A glass transition is a transition from a hard and brittle state to a more viscous state. One such glass, called silicate glass, develops from collections of SiO_4 molecules. This material has three states, a crystalline state which corresponds to a regular array of atoms, an amorphous state which follows the glass transition, and a liquid state. Hiraoka et al. (2016) used molecular dynamics simulations to create examples of collections of atoms in each of these states. The resulting persistence diagrams are shown in Figure 6.29.

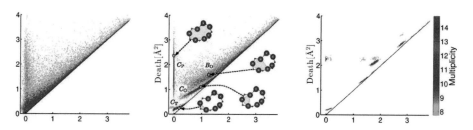

Figure 6.29 Persistence diagrams for various states of silicate glass; from Hiraoka et al. (2016).

We note first that in the crystalline state, the persistence diagram is concentrated at local points, which might be thought of as point masses. In the liquid state, there is a fairly regular distribution near the y-axis and the line $y = x$. In the amorphous state, though, there are three curves of concentration, denoted by C_P, C_T, and C_O in the middle diagram, as well as a point concentration denoted by B_O. Hiraoka et al. (2016) also determined the provenance of these curves as coming from deformations of a ring consisting of seven nodes into situations where various pairs of atoms (red discs) become sufficiently close to each other that they create a new, shorter loop, partially filling in the void in the middle of the ring of atoms. The pictures in the middle figure indicate the various deformations that occur: the green areas show configurations of how the void can be reduced. They also mark where the corresponding configuration shows up in the persistence diagram.

In addition to distinguishing between the states and the explanation of what causes the distinctions between their barcodes, Hiraoka et al. (2016) were also able to gain understanding of other properties of the materials. For example, they were able to determine the wavelength of the so-called *first sharp diffraction peak* from the barcode description. They also verified that the curve structure in the amorphous state was preserved under strain, which is an indication that the persistence diagrams are able to properly encode the elastic properties of the material.

This case study as well as the case study in Section 6.8 demonstrate a very interesting and powerful capability of persistence diagrams, that of detecting small-scale structures that are somewhat akin to the texture within point clouds. The point clouds that are considered here and in the study in Section 6.8 can both be viewed as sampled from $\mathbb{R}^3$, which is a contractible space. One therefore does not obtain any large-scale structure such as that found in the case study in Section 6.1. However, the homology has the ability to distinguish between the different lower-level structures that arise in amorphous materials and in the cosmic web. These applications of topological methods are more akin to the subject referred to as *shape theory* (see Dydak & Segal 1978, or Borsuk 1975), which concerns itself with the study of locally complicated point-set topological objects such as the *Warsaw circle* or the *Sierpinski gasket*.

6.11   Infectious Diseases

The problem of studying the progression of infectious diseases clearly has a great deal of importance to anyone concerned with public health and the treatment of persons infected by disease. It would be very useful to build models that describe where a patient is within the progression of the disease, and what the characteristic behaviors of measurable quantities such as physiological variables, as well as genomic measures such as gene expression, are for the various stages of the disease. An important desideratum for such a model is that it should not depend explicitly on time, but should rather concern itself exclusively with the state of the subject. One reason is that it has been observed that not all subjects traverse a disease trajectory at the same rate, and they may even increase their rate through the trajectory in some some parts of it and decrease it in others. A second reason is that unless we are in laboratory circumstances, we do not generally have information about the time of infection. We might also like to have the ability to predict, based on a subject's state, whether the subject will recover or not.

It is useful to consider what such a model might look like. It would be natural to consider a Mapper model (introduced in Section 4.3.5) of a data set constructed from some physiological variables and perhaps some genomic variables. What we would now expect from the Mapper model is the following. We begin in a healthy state, and as the infection occurs and begins to progress we traverse a trajectory moving away from the

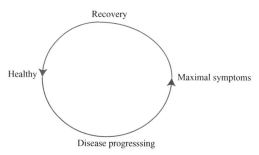

Figure 6.30 Progression of disease

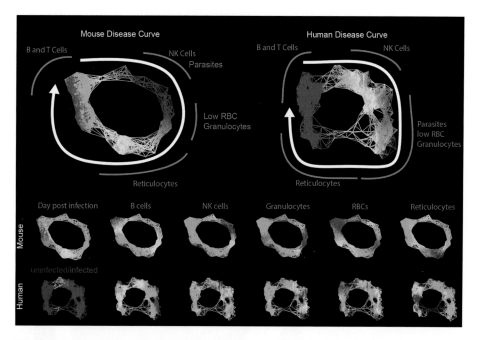

Figure 6.31 Disease maps constructed in Mapper for mice (upper left) and humans (upper right) suffering from malaria. Different phases of the cycles (lower diagrams) are characterized by a prevalence of different types of cells. From Torres et al. (2016), published under CC BY 4.0.

healthy state. Ultimately we arrive at a kind of maximal disease state, but at that point the immune system likely has activated and we begin the return to the healthy state. The key observation, though, is that the return trajectory does not simply retrace the disease trajectory but moves along a different trajectory, eventually arriving at the healthy state. A schematic picture is given in Figure 6.30.

The above discussion suggests that the Mapper model, which we will refer to as a *disease map*, should take the form of a loop.

In Torres et al. (2016) and Louie et al. (2016), such Mapper models are constructed for malaria in humans and in mice and for Listeria in fruit flies, respectively.

Torres et al. (2016) constructed a Mapper model of data sets obtained from blood samples taken daily from mice which had been infected in the laboratory with the malaria parasite *Plasmodium chabaudi*, and from humans suffering from malaria. The blood samples were used to perform microarray studies. Microarrays are a technology that assigns to a blood sample a long vector, whose coordinates correspond to genes and where the actual entries are so-called expression levels within the sample for the given gene. In the case of the mouse study described here, the microarray study produced expression levels for 109 genes.

In Torres et al. (2016), it was shown that various combinations of variable pairs (for example the red blood cell count and the so-called natural killer cell counts) could produce a loop when time was considered, but that standard correlation analysis would not show this result. If one had a patient where one had a complete time series of data, one might be able to obtain useful information. However, what Torres et al. wanted was a method that did not require the availability of longitudinal data for the sample, and they therefore embarked on a full multidimensional analysis based on the 109 features, using the Mapper method. Once produced, this disease map gives a model which is easy to interrogate, and which will permit the placement of a new individual in the map to see where it lies in the progression of the disease; see Figure 6.31.

In Louie et al. (2016), a similar analysis was performed using fruit flies infected with the pathogenic bacteria *Listeria monocytogenes*. The resulting Mapper analysis produces the loop shown in Figure 6.32.

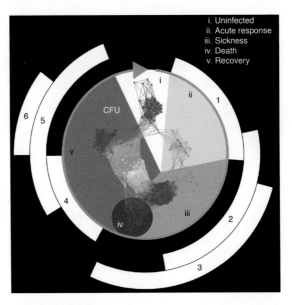

Figure 6.32 Mapper model for fruit flies infected with *Listeria monocytogenes*. The green circular arrow indicates the progression of the disease. From Louie et al. (2010), published under CC BY 4.0.

In addition to the loopy structure, one observes a spur colored in red at the bottom of the loop. It turns out that if a particular infected fly actually lies in this spur, it is very unlikely to proceed around the loop, and is therefore likely not to survive. This kind of feature is difficult to discover using standard time series analysis.

References

Abergel, A. et al. (2011), "Planck early results. XXIV. Dust in the diffuse interstellar medium and the galactic halo", *Astronomy and Astrophysics*, **536**.

Adams, H. & Carlsson, G. (2015), "Evasion paths in mobile sensor networks", *International Journal of Robotics Research* **34**(1), 90–104.

Adams, H., Emerson, T., Kirby, M., Neville, R., Peterson, C., Shipman, P. et al. (2017), "Persistence images: a stable vector representation of persistent homology", *Journal of Machine Learning Research* **18**(1), 218–252.

Adler, R. (1981), *The Geometry of Random Fields*, Wiley Series in Probability and Mathematical Statistics, John Wiley & Sons.

Adler, R. & Taylor, J. (2007), *Random Fields and Geometry*, Springer Monographs in Mathematics, Springer.

Adler, R., Bobrowski, O., Borman, M., Subag, E. & Weinberger, S. (2010), "Persistent homology for random fields and complexes", in *Borrowing Strength: Theory Powering Applications. A Festschrift for Lawrence D. Brown*, J. O. Berger, T. T. Cai, and I. M. Johnstone eds., IMS Collections Vol. 6, Institute of Mathematical Science.

Aggarwal, C. (2018), *Neural Networks and Deep Learning*, Springer.

Artin, M. & Mazur, B. (1986), *Étale Homotopy*, Springer.

Baez, J. (2010), "This week's finds in mathematical physics". `http://math.ucr.edu/home/baez/week293.html`

Bank, P. D. (1971), "Protein data bank", *Nature New Biology* **233**, 223.

Bardeen, J., Bond, J., Kaiser, N. & Szalay, A. (1986), "The statistics of peaks of gaussian random fields", *Astrophysical Journal* **304**, 15–61.

Bauer, U., Kerber, M., Reininghaus, J. & Wagner, H. (2017), "Phat – persistent homology algorithms toolbox", *Journal of Symbolic Computation* **78**, 76–90.

Bennett, C., Hill, R., Hinshaw, G., Nolta, M., Odegard, N., Page, L. et al., (2003), "First-year wilkinson microwave anisotropy probe (wmap)* observations: foreground emission", *Astrophysical Journal Supplement Series* **148**(1), 97–117.

Berwald, J., Gidea, M. & Vejdemo-Johansson, M. (2014), "Automatic recognition and tagging of topologically different regimes in dynamical systems", *Discontinuity, Nonlinearity and Complexity* **3**(4), 413–426.

Biscio, C. & Møller, J. (2016), 'The accumulated persistence function, a new useful functional summary statistic for topological data analysis, with a view to brain artery trees and spatial point process applications', arXiv preprint arXiv:1611.00630.

Bishop, R. & Crittenden, R. (1964), *Geometry of Manifolds*, Academic Press.

Blumberg, A., Gal, I., Mandell, M. & Pancia, M. (2014), "Robust statistics, hypothesis testing, and confidence intervals for persistent homology on metric measure spaces", *Foundations of Computational Mathematics* **14**(4), 745–789.

Bonnet, O. (1848), "Memoire sur la theorie generale des surfaces", *Journal de l'École polytechnique* **19**, pp. 1–146.

Borg, I. & Groenen, P. (1997), *Modern Multidimensional Scaling: Theory and Applications*, Springer.

Borsuk, K. (1948), "On the imbedding of systems of compacta in simplicial complexes", *Fundamenta Mathematicae* **35**(1), 217–234.

Borsuk, K. (1975), *Shape Theory*, PWN-Polish Scientific Publishers.

Bottou, L., Cortes, C., Denker, J., Drucker, H., Guyon, I., Jackel, L. et al., (1994), "Comparison of classifier methods: a case study in handwritten digit recognition", in *Pattern Recognition, 1994,* Vol. 2, Conference B: Computer Vision & Image Processing *Proceedings of the 12th International IAPR Conference*, IEEE, pp. 77–82.

Bronstein, A. M., Bronstein, M. M. & Kimmel, R. (2008), *Numerical Geometry of Non-Rigid Shapes*, Springer Science & Business Media.

Bubenik, P. (2015), "Statistical topological data analysis using persistence landscapes", *Journal of Machine Learning Research* **16**(1), 77–102.

Buneman, P. (1974), "A note on the metric properties of trees", *Journal of Combinatorial Theory B* **17**, 48–50.

Burago, D., Burago, Y. & Ivanov, S. (2001), *A Course in Metric Geometry*, Vol. 33 of Graduate Studies in Mathematics, American Mathematical Society.

Busaryev, O., Dey, T., Sun, J. & Wang, Y. (2010), "ShortLoop software for computing loops in a shortest homology basis", Software, `https://web.cse.ohio-state.edu/~dey.8/shortloop.html`.

Cang, Z. & Wei, G. (2017), 'Topology based deep convolutional and multi-task neural networks for biomolecular property predictions', *PLoS Computational Biology* **13**(7).

Cang, Z. & Wei, G. (2018), 'Integration of element specific persistent homology and machine learning for protein-ligand binding affinity prediction', *International Journal for Numerical Methods in Biomedical Engineering* **34**(2), e2914.

Cang, Z. & Wei, G. (2020), 'Persistent cohomology for data with multicomponent heterogeneous information', *SIAM Journal Mathematics of Data Science* **2**(2), 396–418.

Cang, Z., Mu, L. & Wei, G. (2018), 'Representability of algebraic topology for biomolecules in machine learning based scoring and virtual screening', *PLoS Computational Biology* **14**(1).

Carlsson, G. & de Silva, V. (2010), "Zigzag persistence", *Foundations of Computational Mathematics* **10**(4), 367–405.

Carlsson, G. & Filippenko, B. (2020), "The space of sections of a smooth function", *arXiv preprint: arXiv:2006.12023*.

Carlsson, G. & Gabrielsson, R. (2020), "Topological approaches to deep learning". In *Topological Data Analysis, Proceedings of Abel Symposium 2018*, Springer, pp. 119–146.

Carlsson, G. & Kališnik Verovšek, S. (2016), "Symmetric and r-symmetric tropical polynomials and rational functions", *Journal of Pure and Applied Algebra* **220**, 3610–3627.

Carlsson, G. & Zomorodian, A. (2009), "The theory of multidimensional persistence", *Discrete and Computational Geometry* **42**(1), 71–93.

Carlsson, G., Zomorodian, A., Collins, A. and Guibas, L. (2005a), "Persistence barcodes for shapes", in *Proceedings of the 2004 Eurographics/ACM SIGGRAPH Symposium on Geometry Processing*, ACM, pp. 124–135.

Carlsson, G., Zomorodian, A., Collins, A. & Guibas, L. (2005b), "Persistence barcodes for shapes", *International Journal of Shape Modeling* **11**(2), 149–187.

Carlsson, G., Ishkhanov, T., de Silva, V. & Zomorodian, A. (2008), "On the local behavior of spaces of natural images", *International Journal of Computer Vision* **76**(1), 1–12.

Carlsson, G., de Silva, V. & Morozov, D. (2009), "Zigzag persistent homology and real-valued functions", in *Proceedings of the 25th Annual Symposium on Computational Geometry* ACM.

Carrière, M. & Oudot, S. (2018), 'Structure and stability of the one-dimensional mapper', *Foundations of Computational Mathematics* **18**(6), 1333–1396.

Čech, E. (1932), "Höherdimensionale Homotopiegruppen", *Verhandlungen des Internationalen Mathematikerkongress* **2**, 203.

Chan, J., Carlsson, G. & Rabadan, R. (2013), "The topology of viral evolution", *Proceedings of the National Academy of Sciences* **110**(46), 18566–18571.

Chatterjee, S. (2008), *Crystallography and the World of Symmetry*, Springer.

Chazal, F., Cohen-Steiner, D., Guibas, L., Memoli, F. & Oudot, S. (2009), "Gromov–Hausdorff stable signatures for shapes using persistence", in *Eurographics Symposium on Geometry Processing 2009*, Vol. 28.

Chazal, F., Cohen-Steiner, D. & Merigot, Q. (2011), "Geometric inference for probability measures", *Foundations of Computational Mathematics* **11**(6), 733–751.

Chazal, F., de Silva, V., Glisse, M. & Oudot, S. (2016), *The Structure and Stability of Persistence Modules*, Springer.

Chung, Y.-M. & Lawson, A. (2019), 'Persistence curves: a canonical framework for summarizing persistence diagrams', *arXiv preprint arXiv:1904.07768* .

Chung, Y.-M., Hu, C.-S., Lawson, A. & Smyth, C. (2019), "TopoResNet: a hybrid deep learning architecture and its application to skin lesion classification", *arXiv preprint arXiv:1905.08607* .

CMU Graphics Lab (2012), "CMU graphics lab motion capture database", http://mocap.cs.cmu.edu/, accessed November 2012.

Cohen-Steiner, D., Edelsbrunner, H. & Harer, J. (2007), "Stability of persistence diagrams", *Discrete and Computational Geometry* **37**(1), 103–120.

Cohen-Steiner, D., Edelsbrunner, H., Harer, J. & Mileyko, Y. (2010), "Lipschitz functions have l^p-stable persistence", *Foundations of Computational Mathematics* **10**(2), 127–139.

Collins, A., Zomorodian, A., Carlsson, G. & Guibas, L. (2004), "A barcode shape descriptor for curve point cloud data", *Computers & Graphics* **28**(6), 881–894.

Curry, J. (2013), "Sheaves, cosheaves and applications", arXiv E-Print 1303.3255.

Dalbec, J. (1999), "Multisymmetric functions", *Beiträge Algebra Geometrie.* **40**(1), 27–51.

de Silva, V. & Carlsson, G. (2004), "Topological estimation using witness complexes", in M. Alexa & S. Rusinkiewicz, eds. *Proceedings of the Eurographics Symposium on Point-Based Graphics*.

de Silva, V. & Ghrist, R. (2006), "Coordinate-free coverage in sensor networks with controlled boundaries via homology", *International Journal of Robotics Research* **25**(12), 1205–1222.

de Silva, V., Morozov, D. & Vejdemo-Johansson, M. (2011), "Persistent cohomology and circular coordinates", *Discrete and Computational Geometry* **45**(4), 737–759.

de Silva, V., Skraba, P. & Vejdemo-Johansson, M. (2012), "Topological analysis of recurrent systems", in *Proceedings of the Workshop on Algebraic Topology and Machine Learning, NIPS*

de Silva, V., Munch, E. & Patel, A. (2016), "Categorified reeb graphs", *Discrete and Computational Geometry* 854–906.

Derksen, H. & Weyman, J. (2005), 'Quiver representations', *Notices of the American Mathematical Society* **52**(2), 200–206.

Dey, T. K., Sun, J. & Wang, Y. (2011), "Approximating cycles in a shortest basis of the first homology group from point data", *Inverse Problems* **27**(12), 124004.

Diaconis, P. & Friedman, J. (1980), "*M* and *n* plots", Technical report, Department of Statistics, Stanford University.

Donaldson, S. (1984), "Gauge theory and topology", in *Proceedings of the International Congress of Mathematicians* pp. 641–645.

Donoho, D. (1999), "Wedgelets: nearly minimax estimation of edges", *Annals of Statistics* **27**, 859–897.

Doolittle, W. (1999), "Phylogenetic classification and the universal tree", *Science* **284**(5423), 2124–2129.

Drummond, A., Nicholls, F., Rodrigo, A. & Solomon, W. (2002), "Estimating mutation parameters, population history, and genealogy simultaneously from temporally spaced sequence data", *Genetics* **161**(3), 1307–1320.

Dummit, D. & Foote, R. (2004), *Abstract Algebra*, John Wiley and Sons.

Duponchel, L. (2018 (. "Exploring hyperspectral imaging data sets with topological data analysis", *Analytica Chimica Acta* **1000**, 123–131.

Duvroye, L. (1987), *A Course in Density Estimation*, Birkhäuser.

Duy, T., Hiraoka, Y. & Shirai, T. (2016), "Limit theorems for persistence diagrams", arXiv preprint arXiv:1612.08371.

Dydak, J. & Segal, J. (1978), *Shape Theory: An Introduction*, Springer.

Easley, D. & Kleinberg, J. (2011), *Networks, Crowds, and Markets*, Cambridge University Press.

Edelsbrunner, H., Kirkpatrick, D. & Seidel, R. (1983), "On the shape of a set of points in the plane", *IEEE Transactions on Information Theory* **29**(4), 551–559.

Eilenberg, S. (1944), "Singular homology theory", *Annals of Mathematics* **45**(2), 407–447.

Eisenstein, J. (2018), *Introduction to Natural Language Processing*, MIT Press.

Erickson, J. & Whittlesey, K. (2005), "Greedy optimal homotopy and homology generators", in *Proceedings of the 16th Annual ACM–SIAM Symposium on Discrete Algorithms*, pp. 1038–1046.

Euler, L. (1741), "Solutio problematis ad geometriam situs pertinentis", **8**, 128–140.

Euler, L. (1752a), "Elementa doctrinae solidorum", *Novi Comm. Acad. Sci. Imp. Petropol.* (4), 109–140. (*Opera Omnia Series 1*, vol. 26, pp. 71–93.)

Euler, L. (1752b), "Demonstratio nonnullarum insignium proprietatum quibas solida hedris planis inclusa sunt praedita", *Novi Comm. Acad. Sci. Imp. Petropol.* (4) 140–160. (*Opera Omnia Series 1*, vol. 26, pp. 94–108.)

Everitt, B., Landau, S., Leese, M. & Stahl, D. (2011), *Cluster Analysis*, John Wiley and Sons.

Faúndez-Abans, M., Ormenoño, M. & de Oliveira-Abans, M. (1996), "Classification of planetary nebulae by cluster analysis and artificial neural networks", *Astronomy and Astrophysics Supplement* **116**, 395–402.

Federer, H. (1969), *Geometric Measure Theory*, Springer.

Feichtinger, H. & Strohmer, T. (1998), *Gabor Analysis and Algorithms: Theory and Applications*, Birkhäuser.

Felsenstein, J. (2004), *Inferring Phylogenies*, Sinauer Associates.

Fréchet, M. (1944), "L'intégrale abstraite d'une fonction abstraite d'une variable abstraite et son application á la moyenne d'un élément aléatoire de natur quelconque", *Reviews of Science* **82**, 483–512.

Fréchet, M. (1948), "Les éléments aléatoires de nature quelconque dans un espace distancié", *Annals Institut Henri Poincaré* **82**, 215–310.

Freitag, E. & Kiehl, R. (1988), *Étale Cohomology and the Weil Conjecture*, Springer.

Gabriel, P. (1972), "Unzerlegbare Darstellungen I", *Manuscr. Math.* (6), 71–103.

Ghrist, R. & Krishnan, S. (2017), 'Positive Alexander duality for pursuit and evasion', *SIAM Journal on Applied Algebra and Geometry* **1**(1), 308–327.

Goodfellow, I., Bengio, Y. & Courville, A. (2016), *Deep Learning*, MIT Press.

Gott, J., Dickinson, M. & Melott, A. (1986), "The sponge-like topology of large-scale structure in the universe", *Astrophysical Journal* **306**, 341–357.

Greven, A., Pfaffelhuber, P. & Winter, A. (2009), "Convergence in distribution of random metric measure spaces", *Prob. Theo. Rel. Fields* **145**(1), 285–322.

Grosman, L., Smikt, O. & Smilansky, U. (2008), "On the application of 3-d scanning technology for the documentation and typology of lithic artifacts", *Journal of Archaeological Science* **35**(12), 3101–3110. `www.sciencedirect.com/science/article/pii/S0305440308001398`

Gross, P. & Kotiuga, P. (2004), *Electromagnetic Theory and Computation – A Topological Approach*, I Mathematical Sciences Research Institute Publications Vol. 48, Cambridge University Press.

Gusfield, D. (1997), *Algorithms on Strings, Trees, and Sequences*, Cambridge University Press.

Hamilton, A., Gott, J. & Weinberg, D. (1986), "The topology of the large-scale structure of the universe", *Astrophysical Journal* (309), 1–12.

Hartigan, J. (1975), *Clustering Algorithms*, Wiley Series in Probability and Mathematical Statistics, John Wiley and Sons.

Hartshorne, R. (1977), *Algebraic Geometry*, Springer.

Hastie, T., Tibshirani, R. & Friedman, J. (2009), *The Elements of Statistical Learning: Data Mining, Inference, and Prediction*, Springer.

Hatcher, A. (2002), *Algebraic Topology*, Cambridge University Press.

He, K., Zhang, X., Ren, S. & Sun, J. (2016), "Deep residual learning for image recognition", in *Proceedings of the IEEE Conference on Computer Vision and Pattern Recognition* pp. 770–778.

Hein, J. (2003), *Discrete Mathematics*, Jones and Bartlett.

Hilton, P. (1988), "A brief, subjective history of homology and homotopy theory in this century", *Mathematics Magazine* **60**, 282–291.

Hiraoka, Y., Nakamura, T., Hirata, A., Escolar, E., Matsue, K. & Nishiura, Y. (2016), "Hierarchical structures of amorphous solids characterized by persistent homology", in *Proceedings of the National Academy of Sciences of the United States of America* , Vol. 13, pp. 7035–7040.

Horst, A. M., Hill, A. P. & Gorman, K. B. (2020), "Palmerpenguins: Palmer Archipelago (Antarctica) penguin data." R package version 0.1.0. `https://allisonhorst.github.io/palmerpenguins/`

Hubel, D. (1988), *Eye, Brain, and Vision*, Holt, Henry, and Co.

Hubel, D. H. & Wiesel, T. N. (1964), "Effects of monocular deprivation in kittens", *Naunyn-Schmiedebergs Archiv for Experimentelle Pathologie und Pharmakologie* **248**, 492–497. `http://hubel.med.harvard.edu/papers/HubelWiesel1964NaunynSchmiedebergsArchExpPatholPharmakol.pdf`. (Accessed September 10, 2017)

Huchra, J., Jarrett, T., Skrutskie, M., Cutri, R., Schneider, S. & Macri, L. (2005), "The 2mass redshift survey and low galactic latitude large-scale structure", in *Nearby Large-Scale Structures and the Zone of Avoidance, Proceedings of the 2004 Conference in Cape Town, South Africa*, Vol. 329 of ASP Conference Series.

Hurewicz, W. (1935), "Beiträge zur Topologie der Deformationen. i. Höherdimensionale Homotopiegruppen", *Proceedings of Koninklijke Academie Wetenschappen, Amsterdam* **38**, 112–119.

Kališnik, S. (2019), "Tropical coordinates on the space of persistence barcodes", *Foundations of Computational Mathematics* **19**(1), 101–129.

Kantz, H. & Schreiber, R. (2004), *Nonlinear Time Series Analysis*, Cambridge University Press.

Karcher, H. (1977), "Riemannian center of mass and mollifier smoothing", *Communications in Pure and Applied Mathematics* **30**, 509–541.

Kendall, W. (1990), "Probability, convexity, and harmonic maps with small image: I. uniqueness and fine existence", *Proceedings of the London Mathematical Society (Third Series)* **61**, 371–406.

Khryashchev, D., Chu, J., Vejdemo-Johansson, M. & Ji, P. (2020), "A distributed approach to the evasion problem", *Algorithms* **13**(6), 149.

Kirchgässner, G. & Wolters, J. (2007), *Introduction to Modern Time Series Analysis*, Springer.

Klee, V. (1980), "On the complexity of d-dimensional voronoi diagrams", *Archiv der Mathematik* **34**, 75–80.

Kogan, J. (2007), *Introduction to Clustering Large and High-Dimensional Data*, Cambridge University Press.

Kovacev-Nikolic, V., Bubenik, P., Nikolić, D. & Heo, G. (2016), "Using persistent homology and dynamical distances to analyze protein binding", *Statistical Applications in Genetics and Molecular Biology* **15**(1), 19–38.

Lee, A., Pedersen, K. & Mumford, D. (2003), "The nonlinear statistics of high-contrast patches in natural images", *International Journal of Computer Vision* **1/2/3**(54), 83–103.

Lesnick, M., Rabadan, R. & Rosenbloom, D. (2019), "Quantifying genetic innovation: mathematical foundations for the topological study of reticulate evolution", *SIAM Journal of Applied Algebra and Geometry*, 141–184.

Li, L. et al. (2015). "Identification of type 2 diabetes subgroups through topological analysis of patient similarity." *Science Translational Medicine* **7**

Lipsky, D., Skraba, P. & Vejdemo-Johansson, M. (2011), "A spectral sequence for parallelized persistence", *arXiv preprint arXiv:1112.1245*.

Listing, J. (1848), *Vorstudien zur Topologie*, Vandenhoeck and Ruprecht.

Louie, A., Song, K. H., Hotson, A., Thomas Tate, A. & Schneider, D. S. (2016), "How many parameters does it take to describe disease tolerance?", *PLoS Biology* **14**(4), e1002435.

Love, E., Filippenko, B., Maroulas, V. & Carlsson, G. (2020), "Topological convolutional neural networks", in *Proceedings of the Neurrips Workshop on Topological Data Analysis and Beyond*.

Lum, P., Singh, G., Carlsson, J., Lehman, A., Ishkhanov, T., Vejdemo-Johansson, M. et al. (2013), "Extracting insights from the shape of complex data using topology", *Scientific Reports* **3**, 1236.

Maclagan. D. & Sturmfels, B. (2015), *Introduction to Tropical Geometry*, American Mathematical Society.

MacLane, S. (1998), *Categories for the Working Mathematician*, second edition, Springer-Verlag.

Maleki, A., Shahram, M. & Carlsson, G. (2008), "Near optimal coder for image geometries", in *Proceedings of the IEEE International Conference on Image Processing (ICIP), San Diego, CA*.

Manning, C. & Schütze, H. (1999), *Foundations of Statistical Natural Language Processing*, MIT Press.

Marzinotto, A., Stork, J.A., Dimarogonas, D. V. & Kragic, D. (2014), "Cooperative grasping through topological object representation", in *Proceedings of the 14th IEEE–RAS International Conference on Humanoid Robots* pp. 685–692.

Mileyko, Y., Mukherjee, S. & Harer, J. (2011), "Probability measures on the space of persistence diagrams", *Inverse Problems* **27**, 1–22.

Milnor, J. (1963), *Morse Theory*, Princeton University Press.

Milnor, J. & Stasheff, J. (1974), *Characteristic Classes*, Princeton University Press.

Monro, G. (1987), "The concept of multiset", *Zeitschrift für Mathematische Logik und Grundlagen der Mathematik* (33), 171–178.

Montgomery, D., Peck, E. & Vining, G. (2006), *Introduction to Linear Regression Analysis*, John Wiley Sons.

Munkres, J. (1975), *Topology: a First Course*, Prentice-Hall.

Nelson, B. (2020), "Parameterized topological data analysis", Ph.D. thesis, Stanford University.

Nguyen, D., Gao, K., Chen, J., Wang, R. & Wei, G. (2020a), "Potentially highly potent drugs for 2019-ncov", bioRxiv https://doi.org/10.1101/2020.02.05.936013.

Nguyen, D., Gao, K., Wang, M. & Wei, G. (2020b), "Mathdl: mathematical deep learning for d3r grand challenge 4", *Journal of Computer-Aided Molecular Design* **34**, 131–147.

Nicolau, M., Levine, A. & Carlsson, G. (2011), "Topology based data analysis identifies a subgroup of breast cancers with a unique mutational profile and excellent survival", *Proceedings of the National Aacademy of Science* **108**(17), 7265–7270.

Niyogi, P., Smale, S. & Weinberger, S. (2008), "Finding the homology of submanifolds with high confidence from random samples", *Discrete and Computational Geometry* **39**, 419–441.

Offroy, M. & Duponchel, L. (2016), 'Topological data analysis: a promising big data exploration tool in biology, analytical chemistry, and physical chemistry', *Analytica Chimica Acta* **910**, 1–11.

Otter, N., Porter, M. A., Tillmann, U., Grindrod, P. & Harrington, H. A. (2017), "A roadmap for the computation of persistent homology", *EPJ Data Science* **6**(1), 17.

Park, C., Pranav, P., Chingangram, P., van de Weygaert, R., Jones, B., Vegter, G. et al. (2013), "Betti numbers of gaussian fields", *Journal of the Korean Astronomical Society* **46**, 125–131.

Pedregosa, F., Varoquaux, G., Gramfort, A., Michel, V., Thirion, B., Grisel, O. et al. (2011), "Scikit-learn: machine learning in Python", *Journal of Machine Learning Research* **12**, 2825–2830.

Perea, J. & Carlsson, G. (2014), 'A Klein bottle-based dictionary for texture representation', *International Journal of Computer Vision* **107**(1), 75–97.

Perea, J. & Harer, J. (2015), 'Sliding windows and persistence: an application of topological methods to signal analysis', *Foundations of Computational Mathematics* **15**(3), 799–838.

Poincaré, H. (1895), "Analysis situs", *Journal de l'École Polytechnique* **1**(2), 1–123.

Pokorny, F. T., Stork, J. A. & Kragic, D. (2013), "Grasping objects with holes: a topological approach", in *Proceedings of the 2013 IEEE International Conference on Robotics and Automation*, pp. 1100–1107.

Rabadan, R. & Blumberg, A. (2019), *Topological Data Analysis for Genomics and Evolution*, Cambridge University Press.

Reaven, G. & Miller, R. (1979), 'An attempt to define the nature of chemical diabetes using a multidimensional analysis', *Diabetologia* **16**(1), 17–24.

Reeb, G. (1946), "Sur les points singuliers d'une forme de pfaff completement integrable ou d'une fonction numerique", *Comptes Rendus des Seances de l'Academie des Sciences* **222**, 847–849.

Richardson, E. & Werman, M. (2014), "Efficient classification using the euler characteristic", *Pattern Recognition Letters* **49**, 99–106. `www.sciencedirect.com/science/article/pii/S0167865514002050`

Riehl, E. (2017), *Category Theory in Context*, Courier Dover Publications.

Riemann, B. (1851), "Grundlagen für eine allgemeine Theorie der Functionen einer veränderlichen complexen Grösse", Ph.D. thesis, Georg-August-Universität Göttingen.

Robins, V. (1999), "Towards computing homology from finite approximations", *Topology Proceedings* **24**(1), 503–532.

Robinson, M. (2014), *Topological Signal Processing*, Springer.

Rolfsen, D. (1976), *Knots and Links*, Publish or Perish.

Saggar, M. et al. (2018). "Towards a new approach to reveal dynamical organization of the brain using topological data analysis." *Nature Communications* **9**.

Scott, D. (2015), *Multivariate Density Estimation. Theory, Practice, and Visualization*, John Wiley and Sons.

Segal, G. (1968), "Classifying spaces and spectral sequences", *Publications Mathématiques de l'IHÉS* **34**, 105–112.

Singh, G., Mémoli, F. & Carlsson, G. (2007), "Topological methods for the analysis of high dimensional data sets and 3d object recognition", in *Proceedings of the Conferences on Point Based Graphics 2007*.

Skryzalin, J. & Carlsson, G. (2017), "Numeric invariants from multidimensional persistence", *Journal of Applied and Computational Topology* **1**(1), 89–119.

Sneath, P. & Sokal, R. (1973), *Numerical Taxonomy: The Principles and Practice of Numerical Classification*, W. H. Freeman and Co.

Sousbie, T. (2011), "The persistent cosmic web and its filamentary structure – I. Theory and implementation", *Monthly Notices of the Royal Astronomical Society* **414**, 350–383.

Sousbie, T., Pichon, C. & Kawahara, H. (2011), "The persistent cosmic web and its filamentary structure – II. Illustrations", *Monthly Notices of the Royal Astronomical Society* **414**, 384–403.

Spergel, D. N., Bean, R., Doré, O., Nolta, M. R., Bennett, C. L., Dunkley, J. et al. (2007), "Three year Wilkinson Microwave Anisotropy Microwave Probe (WMAP) observations: implications for cosmology." *Astrophysical Journal Supplement Series* **170**(2), 377.

Stork, J. A., Pokorny, F. T. & Kragic, D. (2013), "A topology-based object representation for clasping, latching and hooking", in *Proceedings of thea IEEE–RAS International Conference on Humanoid Robots*, pp. 138–145.

Symons, M. (1981), "Clustering criteria and multivariate normal mixtures", *Biometrics* **37**(1) pp. 35–43.

Szabo, R. (2000), *Equivariant Cohomology and Localization of Path Integrals*, Springer.

Takens, F. (1981), "Detecting strange attractors in turbulence", *Lecture Notes in Mathematics* **898**(1), 366–381.

Tellegen, B. (1952), "A general network theorem, with applications", *Philips Research Reports* **7**, 256–269.

Tenenbaum, J., de Silva, V. & Langford, J. (2000), "A global geometric framework for non-linear dimensionality reduction", *Science* **290**(5500), 2319–2323.

Torres, B., Oliveira, J., Tate, A., Rath, P., Cumnock, K. & Schneider, D. (2016), "Tracking resilience to infections by mapping disease space", *PLoS Biology* **14**.

Tralie, C., Saul, N. & Bar-On, R. (2018), "Ripser.py: a lean persistent homology library for python", *Journal of Open Source Software* **3**(29), 925. https://doi.org/10.21105/joss.00925

Turner, K., Mileyko, Y., Mukherjee, S. & Harer, J. (2014), "Fréchet means for distributions of persistence diagrams", *Discrete and Computational Geometry* **52**(1), 44–70.

van de Weygaert, R., Vegter, G., Edelsbrunner, H., Jones, B., Pranav, P., Park, C. et al. (2011), "Alpha, betti, and the megaparsec universe: on the topology of the cosmic web", in *Transactions on Computational Science XIV*, Lecture Notes in Computer Science, Vol. 6970, pp. 60–101.

van Hateren, J. & van der Schaaf, A. (1990), "Statistical dependence between orientation filter outputs used in a human vision based image code", *Proceedings SPIE Visual Communication and Image Processing* **1360**, 909–922.

Vandermonde, A. (1771), "Remarques sur les problèmes de situation", *Memoires de l'Acadèmie Royale des Sciences*, 566–574.

Vapnik, V. (1998), *Statistical Learning Theory*, John Wiley and Sons.

Vejdemo-Johansson, M. & Leshchenko, A. (2020), "Certified mapper: repeated testing for acyclicity and obstructions to the nerve lemma", in *Topological Data Analysis*, Springer, pp. 491–515.

Vejdemo-Johansson, M., Pokorny, F., Skraba, P. & Kragic, D. (2015), "Cohomological learning of periodic motion", *Applicable Algebra in Engineering, Communication and Computing* **26**(1–2), 5–26.

Wahab, M. (2014), *Essentials of Crystallography*, Narosa Publishing House.

Weininger, D. (1988), "Smiles, a chemical language and information system. 1. Introduction to methodology and encoding rules", *Journal of Chemical Information and Computer Sciences* **28**, pp. 31–36.

Xia, K., Zhao, Z. & Wei, G. (2015), "Multiresolution persistent homology for excessively large biomolecular datasets", *Journal of Chemical Physics* **143**(13), 134103.

Yanai, H., Takeuchi, K. & Takane, Y. (2011), *Projection Matrices, Generalized Inverse Matrices, and Singular Value Decomposition*, Springer.

Zomorodian, A. (2010), "The tidy set: a minimal simplicial set for computing homology of clique complexes", *Proceedings of the 2010 Annual Symposium on Computational Geometry* ACM, pp. 257–266.

Zomorodian, A. & Carlsson, G. (2005), "Computing persistent homology", *Discrete and Computational Geometry* **33**(2), 247–274.

Index

Printed in the United States
by Baker & Taylor Publisher Services